T0074611

THE NEW BIOLOGY

THE NEW BIOLOGY

A Battle between Mechanism and Organicism

MICHAEL J. REISS and MICHAEL RUSE

HARVARD UNIVERSITY PRESS
Cambridge, Massachusetts
London, England
2023

Printed in the United States of America

First printing

Cataloging-in-Publication Data is available from the Library of Congress
ISBN: 978-0-674-97224-7 (alk. paper)

To Fraser Watts

CONTENTS

THE NEW BIOLOGY

PROLOGUE

The Scientific Revolution, lasting from Copernicus to Newton, was above all a change in what linguists call "root metaphors," from seeing the world as an organism—organicism—to seeing the world as a machine—mechanism. To use other language, science pre-Revolution demanded that one think of entities as functioning wholes, "holism." Science post-Revolution worked by looking at entities as composed of individual parts, "reductionism." The flies in the mechanistic ointment were living organisms. They seemed too intricately constructed for us to think that they could be the product of the blind laws we associate with machines. People worried about this problem from the time of Robert Boyle, in the seventeenth century, to Charles Darwin, in the nineteenth century, who claimed that through his mechanism of natural selection, we can explain the nature of organisms using only blind laws. Not all were convinced, and until the end of the nineteenth century, there were many—professional biologists as well as laypeople—who thought that a return to the old metaphor of the organism was necessary. In the first half of the twentieth century, thanks to advances in the study of heredity, culminating in the discovery of the structure of the genetic material, DNA, in 1953, to many, mechanism was all triumphant.

It turned out, however, that it was too soon to write obituaries for organicism. It is true that understanding the way in which DNA functions demanded one think in terms of its parts, reductionism; but many phenomena, most particularly the growth and development of organisms, seemed still to demand a more integrated understanding, holism. Today, there is a lively, often bitter, divide among biologists over this division. It is the aim of this book to throw light on the controversy. We stress that we write not as advocates of one or the other position. For a start, neither of us is any longer a professional scientist. We are, however, educators, and one of us is a philosopher and historian; hence, the task we set ourselves is that of understanding, of giving readers tools whereby they might themselves reach informed conclusions about the relative merits of the two approaches.

To this end, in this book, we discuss such terms as mechanism, reductionism, organicism, and holism. We explore the broader significance of these developments in biology, including their philosophical, educational, religious, and policy relevance. We examine the ongoing debate between mechanism / reductionism and organicism / holism and ask whether we are simply witnessing something of a corrective to an historical anomaly in the history of biology, returning now to a more balanced vision of the discipline in which reductionism and holism both play complementary roles, or if we are at the beginning of the emergence of a "New Biology," in which a unified, holistic understanding of biology will hold sway.

A variety of developments have contributed to recent moves in biology away from the reductionism that sometimes accompanies molecular biology, cell biology, and genetics. One such development has been contemporary understandings of inheritance, which, rather than simply explaining the appearance of organisms, their phenotypes, in terms of their genes, recognize that genes themselves interact and are sensitive to triggers from the environment that can switch them on or off. The interface between evolutionary and developmental biology ("evo-devo") has perhaps been at the epicenter of the New Biology (sticking with this term for the moment), but there are other important developments too. Ecology recognizes the significance for each species of its interactions with other species. More generally, systems biology—meaning that the level of analysis is the system as a whole (e.g., an entire cell or an ecosystem) rather than focusing only on its separated components—recognizes that there are many biological phenomena that cannot be adequately understood in terms of reductionist explanations and is developing mathematical modeling in attempts to capture the complex processes involved. Then there is work on neural plasticity, which recognizes that in addition to structure determining function, it is also possible for function to influence structure. Moving beyond biology itself, the increasing realization throughout the twentieth century (quantum mechanism, chaos theory, etc.) that physics is less deterministic than many have supposed has played a part too.

These are no longer points of deep controversy within the academic biology community. Although, of course, there are localized areas of dispute, as with any science in the process of developing new knowledge, by and large, these points are widely accepted and are guiding current research. A new more systemic, organismal biology *is* gaining ground (Watts & Reiss 2017). However, controversy remains within academic biology and outside of academia; the debate between mechanism/reductionism and organicism/holism is as raucous as it ever was. In the chapters that follow, we discuss the broader significance of these developments in biology and bring them into dialogue with other disciplines. It will become clear that depending where one sits on the mechanism-organicism, reductionism-holism spectrum, there are important implications for how humans see themselves and the world in which we all live.

In his *Metaphysics,* Aristotle said "the totality is not, as it were, a mere heap, but the whole is something besides the parts" (Book VIII, 1045a. 8–10; in Barnes 1984). It is this insight that lies behind the so-called philosophy of "holism" or "organicism," namely that one cannot rest content with a purely reductionistic approach to understanding—particularly the understanding of organisms—but must in some sense look at the whole or the entire body, be this an individual organism or a collection such as a population, species, or even a whole ecosystem. Another popular term is "emergence," meaning that from the parts considered together, new overall properties appear.

An eloquent passage that brings together aspects of both organicism and emergence is found in the writings of John Stuart Mill:

> All organised bodies are composed of parts, similar to those composing inorganic nature, and which have even themselves existed in an inorganic state; but the phenomena of life, which result from the juxtaposition of those parts in a certain manner, bear no analogy to any of the effects which would be produced by the action of the component substances considered as mere physical agents. To whatever degree we might imagine our knowledge

> of the properties of the several ingredients of a living body to be extended and perfected, it is certain that no mere summing up of the separate actions of those elements will ever amount to the action of the living body itself. (Mill [1843] 1974, Book III, Ch. 6, §1)

In the Anglophone world, with the coming of genetics and then of molecular biology, it looked as if reductionism-mechanism had triumphed. Richard Dawkins's *The Selfish Gene* (1976) seemed to be the apotheosis of that philosophy, with the behavior of individual organisms "explained" by the (metaphorical) unconscious urge of their constituent genes to spread at all costs. But the rival organicist philosophy has proved a sturdy plant. In the world of evolution, the eminent population geneticist Sewall Wright (who for many years worshipped with the Unitarians) was always sympathetic to holism. A few years later, a number of scholars, notably Richard Lewontin at Harvard, argued for a more holistic philosophy, first in their more scientific works such as Lewontin's *The Genetic Basis of Evolutionary Change* (1974) and then in more general writings such as Lewontin's *Biology as Ideology: The Doctrine of DNA* (1991), where those espousing a reductionistic account of biology are excoriated as naïve and ill-informed. In paleontology, Stephen Jay Gould argued at length for a more Germanic, organicist view of organisms if we are to understand the diversity of life, first in a massive historical overview, *Ontogeny and Phylogeny* (1977) and then in numerous more scientifically directed publications such as "Darwinism and the expansion of evolutionary theory" (1982).

Most interestingly and pertinently, there has been significant debate about the workings of the main Darwinian mechanism of evolution: natural selection. Like Darwin, most of today's evolutionists think that selection works exclusively or primarily at the level of individuals and their genes (Hertler et al. 2020). But from the beginning, there was always a significant minority who claimed that selection can work at the level of the group, and it is because of this, and only because of this, that we can get genuine altruism, particularly among humans. David Sloan Wilson has been a pioneer in this respect, joined for the past twenty years by the philosopher Elliott Sober. Noteworthy was their *Unto Others: The Evolution and Psychology of Unselfish Behavior* (1998). Ironically, when the definitive work on the evolution of social behavior, *Sociobiology: The*

New Synthesis (1975) by Edward O. Wilson, first appeared, the critics (notably Lewontin and Gould) accused him of being ultra-reductionistic.

A more careful reading would show that this is not necessarily so, and in the years before his death, E. O. Wilson came out emphatically for a group selection approach to evolution (see Wilson & Wilson 2007). Historically, as we shall see, one can link E. O. Wilson back to a powerful group of emergentists at Harvard at the beginning of the twentieth century, including W. M. Wheeler, the ant specialist, Walter B. Cannon, the physiologist, and Lawrence J. Henderson, the biochemist. Sewall Wright, a graduate student at Harvard at the beginning of the twentieth century, also owed major intellectual debts to this circle.

Inheritance segues into the topic of development. From the time of Aristotle, the remarkable phenomenon of development has lent itself to emergent interpretations, as a whole organism apparently miraculously (or "bewitchingly") emerges from seemingly undifferentiated matter. The *Naturphilosophen* (the German Romantics) were particularly interested in this, and embryology became a science of great significance. As Richards (2008) notes in his book on Haeckel, development was embedded in Haeckel's thinking, not least because of his championing of the so-called biogenetic law—ontogeny (development of an individual from single cell to adult) recapitulates phylogeny (the history of the evolution of species).

In the twentieth century, with the coming of genetics, development rather fell by the wayside as organisms were treated something like sausage machines—genes and raw materials in at one end, organisms emerging at the other end. Even embryologists, such as Gavin de Beer (1940), wrote this way to some extent. However, some pushed the significance of embryology in a major way, linking it to a more organicist view of life. Noteworthy were the already mentioned Stephen Jay Gould (1977) and, most particularly, the physician-turned-biologist, very deeply committed to computer programing, Stuart Kauffman (1993, 1995, 2008). Both of these scientists felt that an emergentist philosophy was necessary for a full understanding of the workings of life. Additionally, no account would be complete without mention of the maverick but stimulating thinking of Brian Goodwin. In his *How the Leopard Changed its Spots* (2001), Goodwin argued that genetic reductionism has important limits, and that if we are to understand life, we have to appreciate its capacity to self-organize.

Then came the full flowering of molecular biology with its major insights into the functioning of genes—from DNA to RNA to amino acids to proteins and so on up the chain. Advances in techniques used in molecular and cell biology, such as proteomics (in which increasingly automated approaches are used to study the entire set of proteins produced by a cell or other system) and bioinformatics and computational biology (where software is used to try to make sense of the vast amount of biological information that is increasingly available about organisms), gave added impetus to the hope that by studying the constituents of organisms in more detail, we would be able to understand them.

However, it soon became obvious—and if it was not obvious, then major projects such as the Human Genome Project made it so—that growth is a matter of organization as much as materials, and an emergentist approach was nigh mandatory. This was made clear by the scientists themselves, for instance Sean Carroll in *Endless Forms Most Beautiful: The New Science of Evo Devo* (2005) and plant biologist Ottoline Leyser (Leyser & Wiseman 2020), and in those reflecting on the science, for instance Scott Gilbert (2006). More generally, perspectives tied this thinking to emergentist areas elsewhere in science, for instance in physics in *Complexity and the Arrow of Time* (Lineweaver et al. 2013). In this sense, the move away from "pure reductionism" to a more complex view can be considered to have been, at least for some scientists, a pragmatic response to new data rather than an ideological rejection of the principles of reductionism. The discovery as a result of the Human Genome Project that humans have only about 20,000–25,000 genes that code for proteins rather than the substantially larger number that had been expected (Willyard 2018) played an important part in damping the enthusiastic presumption that once we knew all about the constituents of cells, predicting everything about organisms would flow naturally.

Ecology and environmental issues generally have always attracted those with emergentist leanings. This is hardly surprising because, as historian Gregory Mitman (1992) documented, ecology does push one toward thinking at the macro, even the mega, level. In addition, historical factors are significant for understanding the present-day distribution of organisms. On the one hand, much ecological thinking has been rooted in the "balance of nature" doctrine. Although this had pagan origins, it was taken over by Christian thinkers and pushed people to think holistically. Interestingly and perhaps significantly, Darwin was always cautious about such a balance—it could happen but not necessarily. On the other hand, Herbert Spencer (1820–1903) did see things holistically. Many think that Spencer was the ultimate reductionist, with his coining of the phrase "the survival of the fittest" after he had read Darwin's *On the Origin of Species* (1859). However, this is not the case. His writings on the state as an organism were very influential, especially with the Harvard holists and then later as the University of Chicago grew and flourished. It was not by chance that Chicago (and other places such as Nebraska) became important in the new science of ecology, situated as they were in the Midwest, where environmental change (e.g., the Dust Bowl) was so significant.

Later ecological thinkers, much influenced by G. Evelyn Hutchinson (1948), were more inclined to mechanistic thinking—work on feedback systems in the Second World War was significant here—but some of Hutchinson's most important followers, notably the Odum brothers, were very inclined to holistic thinking. In many cases, this subsequently connected to a sympathy for the brainchild of the English scientist James Lovelock, the Gaia hypothesis, the idea of the Earth as an organism. It is noteworthy that Lovelock's great supporter, Lynn Margulis, was always deeply committed to symbiosis. It is also noteworthy that the Gaia hypothesis is disliked both by scientists such as Richard Dawkins (1982), who think it insufficiently reductionistic, and by evangelical Christians, who think it deifies the Creation (Van Dyke et al. 1996; Ruse 2013).

The philosophical issues surrounding holistic biology are more subtle than is sometimes appreciated. At the very least, the new more organismal biology seems to move away from a mode of explanation that assumes that higher-level phenomena can be explained entirely in terms of lower-level ones. Instead, it moves toward a recognition (i) of the reality and importance of "emergence" (that phenomena that are genuinely new can be seen at higher levels, follow their own laws, and cannot be explained entirely in terms of lower-level phenomena), and (ii) that biological explanations often need to be systemic and to take into account the possibility that lower-level phenomena can be influenced by higher-level organismal factors and by functional context.

There are also issues about determinism to be explored. Epigenetics and other features of contemporary genetics mark a move away from a simplistic genetic determinism in which it is presumed that the phenotype simply follows from the genotype. Of course, that kind of genetic determinism never received much scientific support, existing more in the media rhetoric of a "gene for" this or that, but it nonetheless has had and continues to have a powerful role in the popular (public and school) understanding of biology (Reiss et al. 2020). One issue to be debated is whether holistic biology, with its complex interactionist assumptions, is deterministic and, if so, in what way(s). We therefore explore whether we are dealing with a complex interactive form of determinism or whether holistic biology cannot properly be said to be determinist at all. As with chaos theory, it does not entirely settle the matter that complex systems can be modeled mathematically in a way that makes deterministic assumptions. Both determinism and indeterminism are metaphysical conjectures that become more or less reasonable in the light of a complex network of scientific and philosophical considerations.

We believe that the developments in biology with which this book is concerned have implications for theology and religious belief. The strong reductionism of molecular biology has, in some people's minds, fostered the idea that modern scientific biology is incompatible with religious faith. While that was never a view that stood up to critical examination, work by sociologists (e.g., Ecklund & Johnson 2021) shows it is quite widely accepted, particularly among atheists. The move away from strong reductionism in biology promises to remove what has been, for some people, an obstacle to religious faith, or at best something that sits uneasily with it. There has been fruitful engagement between theology and emergentism (e.g., Gregersen 2017). We see similar scope for theology to engage with the current trend toward holistic biology. There are also constructive theological implications of the New Biology. The systemic complexity of the New Biology points to the interconnectedness of creation in a way that finds a parallel in the religious vision of the unity of all things in God.

Issues about reductionism have been at the heart of work on the interface between science and theology. Philosophical reconciliations have been proposed, including the non-reductive physicalism of Warren Brown and colleagues in *Whatever Happened to the Soul?* (1998), the emergentism of Philip Clayton in *Mind and Emergence* (2006), and elsewhere. These philosophical proposals have been helpful, but we think that they can be strengthened by the scientific developments that we discuss in this book. It is arguable that in the new more organismal biology, science is rescuing itself from the strong reductionist assumptions that have sat uneasily with religious faith. We consider that recent developments in biology make it easier to argue that today's biology is compatible with a theistic faith.

There is a further issue about the relationship between the organicist metaphysics that is sometimes adopted by holistic biology and the mechanistic metaphysics that still predominates in much of biology. The significance of the debate between these two main metaphysical positions in biology is discussed by Michael Ruse in his *Science and Spirituality: Making Room for Faith in an Age of Science* (2010). The debate here is whether holistic biology should be seen as a sophisticated form of mechanistic biology, or whether it is incompatible with mechanistic assumptions and requires an organicist metaphysics. A central question is whether the key scientific findings in the various areas of holistic biology can be interpreted within either theoretical framework, or whether they actually necessitate a nonmechanistic framework.

There is a degree of convergence to be explored between the sense of the interdependence within nature that emerges from the New Biology, the mystical vision of the unity of all things, and the Christian conviction that all things cohere in Christ. This is an approach with a long history, one reflected in fiction and the visual arts (William Blake, Samuel Palmer, Samuel Taylor Coleridge, David Jones, and others) as well as in theology. The New Biology also adds weight to the point often made by Arthur Peacocke, for example in *Paths from Science towards God: The End of All Our Exploring* (2001), that there is "top-down" as well as "bottom-up" causation and that wholes influence parts as well as parts giving rise to wholes. This approach leaves scope for multiple influences of different kinds. For Peacocke, put at its crudest, because this approach is not saying that there is only one kind of causal process, it cannot easily rule anything out, and more readily leaves scope for divine action. We also use the language of bottom-up and top-down thinking without considering any theological implications. Bottom-up thinking entails working in a reductionistic fashion from the parts up to the whole; top-down thinking means seeing things more holistically from the point of view of the whole.

In medicine, a more holistic biology lends strong support to the whole-person approach. To emphasize, we are not against the use of such techniques as molecular biology in medicine. Far from it, we welcome them. Our point is a different one, namely that the (often implicit) presumption made by many that such techniques render redundant consideration of other levels is erroneous, indeed positively harmful. This is true in all sorts of ways. For example, as has been realized for a long time, possession of the gene "for" condition / disease *x* rarely equates straightforwardly to getting condition / disease *x*. A case in point are the various *BRCA* and other genes for breast cancer. This truth has profound implications for genetic counseling, for the allocation of medical resources, for preventive medicine, including so-called preventive (prophylactic) mastectomy, and—what will be something of a common trope—for education, whether at school or post-school level. Another example of the shortcomings of a reductionist approach in medicine is how one should deal with mental health issues, such as schizophrenia, given the increasing realization that such conditions are a result of both internal and external factors; indeed, the right environment can substantially reduce the chances of a person having a schizophrenic attack (World Health Organization 2019a).

More recently, the advent of COVID-19 has clearly indicated that a successful response requires action at every level, from the molecular biology used to identify new variants through to the regulatory and other measures taken at government level with regards to such diverse considerations as mandating masks and social distancing and providing temporary economic support to individuals and businesses adversely affected by the pandemic. At the time of writing, one of the notable features of international comparisons is that many countries have done well at one or more of these levels, but none has done well at all of them.

What, if anything, has the argument we advance in this book to say about human identity—about how we see ourselves in terms of our gender, our ethnicity, and other components of our self? Consider gender. Humans have a tendency to classify and to see the "essence" of things in a way that overstates difference, that draws up rigid boundaries, that essentializes.

So it is with gender. It is easy to presume that humans exist as either males or females and that anything else in an extremely rare aberration. And yet, more careful examination shows that while there is some truth in this assertion, it oversimplifies. This can be for a number of reasons, including chromosomes (not everyone is XX or XY), hormones or, just as importantly, how one feels about oneself. This fact has, as has become clear in recent years, profound implications for a whole range of policy matters, including gender classification in sporting events and issues to do with transgenderism.

Then, consider issues to do with race. In many societies, attitudes toward racial differences are schizoid. On the one hand, both the natural and the social sciences have critiqued (pilloried) naïve attempts (e.g., Herrnstein & Murray 1994) to link racial differences to such characteristics as behavior and educational aptitude (Gould 1981; Donovan et al. 2020). On the other hand, the medical sciences increasingly acknowledge the way in which the racial / ethnic group to which one belongs can correlate, sometimes quite tightly, with one's likelihood of suffering from a whole range of conditions, varying from sickle cell anemia and cystic fibrosis to obesity and type 2 diabetes.

Both human health and the workings of the natural world are more complex than biologists' models sometimes presume. We should therefore always be mindful that an approach that takes seriously a number of levels (from molecules through individual organisms to organisms in ecosystems and beyond) is likely, though more complicated, ultimately to be more fruitful than one that focuses only on one or two of these levels. We also need to remember that there are limits to deterministic predictions. These points do not mean that there is nothing that biology can predict, and throughout, we try to steer a path between overly reductionist and overly holistic approaches.

Finally, we note that the issues we are considering have implications for biology education, which takes place in a number of places and at various times throughout our lives. It happens in our families, in our schools, and through such media as popular science books, Internet posts, natural history museums, and TV and radio. One of the most fundamental issues in biology education is whether one starts with basic scientific principles (e.g., food webs, nutrient cycling, and energy flow in ecology;

genetics and natural selection in evolution; cell biology in physiology) or with real-life instances (e.g., the effects of the extinction and re-introduction of wolves in Yellowstone Park, the evolution of the horse, the regulation of the beating of the heart). There are advantages and disadvantages with either approach, and there is much to be said for learners of biology coming to appreciate that both bottom-up (reductionist) and top-down (more holistic) approaches can provide us with important ways of understanding what is going on.

Equally, the various metaphors we explore for understanding biological systems each have their place. There is, for example, value in seeing organisms as the product of a natural selection that is blind and selfish and allows for no meaning. Indeed, such an understanding of what it is to be human can help strip away layers of flabby self-congratulation. At the same time, there is value in seeing organisms as entities with purpose, a purpose that started, evolutionarily, simply with leaving copies in succeeding generations and, over time, has led to organisms with varying degrees of self-awareness, including to humans capable of appreciating beauty, seeking truth, and striving to be good.

We conclude with a short epilogue that asks whether, overall, resolution of the debate between mechanism / reductionism and organicism / holism that we have been examining requires a New Biology or rather a remolding of existing arguments. In one sense, we don't mind what the answer is. However, we note that biologists still seem polarized about many of the issues we have explored in this book. We speculate that this is partly because of the nature of biology—with a focus that extends in scale from the behavior of quanta of light and of electrons over tiny fractions of a second to the behavior of vast biomes over geological periods of time. It is unsurprising that individual biologists have preferences as to how they work—not everyone likes both baroque and abstract expressionism.

It may therefore not be possible to find a single unified framework for biological explanations that commands universal agreement. What we are clear about, though, is that particularly in the policy implications of biology, there is real danger from too great an emphasis on either mechanism / reductionism or organicism / holism. We need, more than ever, policy to draw on the best of biology and to be sensitive to the ways such knowledge is employed.

Chapter One

MECHANISM TRIUMPHANT

> At all times there used to be a strong tendency among physicists, particularly in England, to form as concrete a picture as possible of the physical reality behind the phenomena, the not directly perceptible cause of that which can be perceived by the senses; they were always looking for hidden mechanisms, and in so doing supposed, without being concerned about this assumption, that these would be essentially the same kind as the simple instruments which men had used from time immemorial to relieve their work, so that a skillful mechanical engineer would be able to imitate the real course of the events taking place in the microcosm in a mechanical model on a larger scale.
>
> —DIJKSTERHUIS 1961, 497

The Scientific Revolution, a time of drastic change in scientific thought, began with Nicolaus Copernicus's positing of a "heliocentric" picture of the universe, with the Sun at the center and all else, including planet Earth, going around it, found in his 1543 publication, *De revolutionibus orbium coelestium* (*On the Revolutions of the Celestial Spheres*). The Revolution can be seen as ending with the 1687 publication of Isaac Newton's *Philosophiæ Naturalis Principia Mathematica* (*Mathematical Principles of Natural Philosophy*), in which Newton shared his laws of motion and of gravitational attraction and explained causally the heliocentric universe and, therefore, our place in it. The Revolution occurred roughly at the same time as the Renaissance—the discoveries of ancient literature that so infused and invigorated fifteenth- and sixteenth-century thinking—and the Reformation—the break from Rome and the rise of Protestantism, fueled by Martin Luther, Jean Calvin, Huldrych Zwingli, and others. These should not be considered as separate isolated movements. There were important interactions, as we see immediately and repeatedly, starting with the appreciation that the key to understanding the Scientific Revolution is as a change of metaphors—from the picture of the world as an organism to that of a machine (Ruse 2021a). To understand how this dramatic intellectual shift occurred, we must follow in the path of the Renaissance and begin with the Greeks (Ruse 2017), the great philosophers Plato and Aristotle, and with their metaphor of organicism.

THE ORGANISM METAPHOR: PLATO

Known as one of the founders of Western philosophy and a pivotal figure in Ancient Greek thought, Plato posited an organic metaphor for life: the universe is orderly and alive, like an organism. This kind of thinking came readily from the nature of Greek life, mainly rural and governed almost entirely by the seasons. First, spring and an awakening, a new birth; next, summer and growth to maturity; after, comes autumn, the reaping of the harvest; and finally, winter, the end of it all—until there comes another new birth, and the pattern repeats itself. What more obvious than to consider the streams and rivers as the lifeblood of the world in which people lived, or the Sun and the rains as feeding this world and encouraging its growth, and then the gradual decline and death that we associate with old age. It was the genius of the Greek philosophers, epitomized by Plato, to start with this world picture, this organic metaphor, and to analyze it, asking about its nature and the implications. Take a moment now and think about an organism—let us say a bird of prey. Once you truly start to consider it, questions begin to formulate. Why does it have feathers? Why does it have wings? Why does it have a beak, and why is the beak curled down? Why are the claws strong and also talons curled? The questions keep coming. Why do some eagles have white feathers on their heads but not on the main body? Why isn't an eagle red like a cardinal? Well, of course, you can readily give answers. The wings are for flying. The beak is for ripping apart small animals of prey. The talons likewise share this purpose. Purpose? Purpose suggests that the features of the bird did not happen by chance. There seems to have been an intelligence at work here—a designer who made sure, for example, that the bird of prey did not have a beak like a sparrow or a thrush or some other bird that lives mainly on seeds and insects.

Most of Plato's writings are in dialogue form, supposedly reporting on spirited discussions led by his mentor Socrates, surrounded as he was with followers (including Plato), eager to understand (Cooper 1997). The early dialogues are considered authentic, but then Plato started to introduce his own ideas, using the historical Socrates as the vehicle. One of the most important, the *Phaedo*—probably more Platonic than Socratic—reports on Socrates's last day on Earth, before he was forced to drink poison. Socrates is being questioned about the existence of God and the afterlife and he gives an early version of what is known as the "argument from design." The world is so intricately formed that it could not be all chance. There must have been a designer—an intelligence with a plan. "One day I heard someone reading, as he said, from a book of Anaxagoras, and saying that it is Mind that directs and is the cause of everything. I was delighted with this cause and it seemed to me to be good, in a way, that Mind should be the cause of all. I thought that if this were so, the directing Mind would direct everything and arrange each thing in the way that was best." So now one has a heuristic, a way to find out about the world. "Then if one wished to know the cause of each thing, why it comes to be or perishes or exists, one had to find what the best way was for it to be, or to be acted upon, or to act" (*Phaedo* 97c–d).

The arguments above apply to birds of prey, but surely they stop when we go beyond organisms? It is here that the organic metaphor starts to kick in—the world as a living organism, going through the phases of birth, of growth, of maturity, and then of decline and death. We see the world in this way, and surely such organization does not happen by chance. Plato offers the ready answer: "Mind directs and is the cause of everything." One of his most influential dialogues, *The Timaeus,* endorsed the vision: the world is an organism. At this point, to make full sense of Plato's philosophical position, we must introduce his theory of Forms. Plato

(1941) posited a kind of rational heaven—eternal, unchanging—with mathematics as a part of it, albeit placed at the bottom. Above mathematics were what Plato called the Forms, templates for ordinary objects of this world. As an example, if I say Dobbin has four legs, I am talking about one specific horse: Dobbin. If I say a horse has four legs, I am talking about the Form of the horse—never changing, outside time and space. Plato's language was that the relationship between the two worlds is that the objects of our world "participate" in the Forms. Forms are the ultimate causal factors. The overall causal factor—the Sun—in our world of change, of becoming, corresponds in the rational world to the Form of the good, or a creator, or a designer—what Plato called the "Demiurge"—the ultimate cause of the other Forms and hence of everything. He argued that the world is an organism, good because it reflects the goodness of the Forms. "Now surely it's clear to all that it was the eternal model he looked at, for, of all the things that have come to be, our universe is the most beautiful and of causes the craftsman is the most excellent. This then is how it has come to be: it is a work of craft, modeled after that which is changeless and is grasped by a rational account, that is, by wisdom" (*Timaeus* 29a).

THE ORGANISM METAPHOR: ARISTOTLE

The second of our great Greek thinkers, Aristotle, was a student of Plato. Aristotle had been a practicing biologist, and his scientific background is apparent when one considers his interest in principles of order (Lennox 2001). While both Plato and Aristotle saw the world as eternal, existing not created, and neither wanted a God interfering in existence, Aristotle wanted no part in an external theory of Forms. He thought that there were determining patterns, but these were inherent in the physical substances of this world rather than in some distinct rational world. He wrote that "if man and such things are substance, that none of the elements in their formulae is the substance of anything, nor does it exist apart from the species or in anything else; I mean, for instance, that no 'animal' exists apart from the particular kinds of animal, nor does any other of the elements present in formulae exist apart" (*Metaphysics* VII, 13). In other words, in Aristotle's view, you have Dobbin and Nelly, two horses. You don't need a third, nonphysical horse, the Form of horse, on top of this. And Dobbin and Nelly are horses. Toby is a dog, not a horse. There is a kind of pattern inherent in Dobbin and Nelly but not in Toby.

Aristotle developed a theory of causality, postulating four kinds of cause to explain the world around us: efficient, material, formal, and final. To introduce the types of knowledge based on the causes, let us consider by way of example a statue of a British foot soldier of the First World War (a "Tommy"). The *efficient* cause—the change humans bring on from before the present state to the present state—is the sculptor or modeler physically making the statue. The *material* cause is what the statue is made from—the bronze or marble. The *formal* cause—showing a link here with Plato—is the picture or conception that the sculptor had in mind. The artist would not, for instance, put a German helmet (a *Pickelhaube*) on a Tommy, but may well give the soldier a rifle. The *final* cause is the reason, intention, or purpose. The statue is intended to be on the village green to honor and remember those who made such sacrifices for us all.

For Aristotle, efficient causation plays a decisive role. When things happen, that is because efficient causation is at work. A bang is heard because something smashed against something else. However, above all, Aristotle is deeply committed to a world governed by final causes. The reason for the bang is the carpenter hammering in nails, building a house for people to live in. In his technical biological work, *The Parts of Animals,* absorbing material causes into efficient causes and formal causes into final causes, he wrote: "The causes concerned in the generation of the works of nature are, as we see, more than one. There is the final cause and there is the motor [efficient] cause." He then stresses that the final cause is more important. Note that, like Plato, he starts with human consciousness and planning, but then moves immediately to the living world, where there is no presumption of mind at work. "Now we must decide which of these two causes [efficient or final] comes first, which second. Plainly, however, that cause is the first which we call the final one. For this is the Reason, and the Reason forms the starting point, alike in the works of art and in works of nature" (*Parts of Animals* 1, 1). Aristotle then goes on to say that it is exactly analogous when we are speaking of organisms. "Now in the works of nature the good end and the final cause is still more dominant than in works of art such as these." Aristotle clearly wants to keep the discussion at the material or the physical. He doesn't want any Forms interfering, and he doesn't want any god—or good—interfering. All the action is down here. In other words, whereas Plato is into what we might call an "external" final-cause governed world, Aristotle is into an "internal" final-cause world. How can this be if there is no outside activity? It must come from within; final cause must involve some kind of motive (presumably, in a sense, efficient) force that drives living things forward to their goals. Now, don't think of this force for Aristotle as a kind of misty presence, a bit like Caspar, the cartoon character. If anything, it is more a principle of ordering. If you are not a Platonist, you still think that the 3-4-5 triangle has a right angle. Somehow, there is a principle of ordering that informs the triangles we draw down here on earth.

THE ORGANISM METAPHOR: THE CHRISTIANS

From the Greeks, we move to the great Christian theologian-philosophers, particularly to St. Augustine and, almost a millennium later, St. Thomas Aquinas. One might think that the followers of this new religion would reject entirely the work of such thinkers as Plato and Aristotle. Even if these thinkers did not fit what today we think of as Pagans—stark-naked Californians dancing around a bonfire (see Chapter 7)—theologically, that is what in truth they were. Plato's Demiurge was a designer not a creator and, although Aristotle posited an "Unmoved Mover," because it was a Perfect Being it spent its time contemplating its own perfection. It had little interest in, perhaps no knowledge of, humans and the world within which we live. Yet, nothing could be further from the truth than hostile Christian indifference to the Greeks. The genius of Augustine and Aquinas and those on whose shoulders they stood was to take the discussion and findings of the great Greek philosophers and use these as philosophical underpinnings of their Christian theology. Augustine was much influenced by the Latin-writing Neoplatonist Plotinus, and Aquinas much influenced by the newly translated (into Latin) Aristotle. For them, as Christians, they obviously had to be Plato-like externalists, identifying the Christian God with Plato's Demiurge—with the addition that their God was not just a designer but a creator too. Both Augustine and Aquinas gave versions of the argument from design. Purpose, final cause, pervades, up to and including the proof of God's existence.

> The fifth way is taken from the governance of the world. We see that things which lack intelligence, such as natural bodies, act for an end, and this is evident from their acting always, or nearly always, in the same way, so as to obtain the best result. Hence it is plain that not fortuitously, but designedly, do they achieve their end. Now whatever lacks intelligence cannot move towards an end, unless it be directed by some being endowed with knowledge and intelligence; as the arrow is shot to its mark by the archer. Therefore some intelligent being exists by whom all natural things are directed to their end; and this being we call God. (Aquinas 1981, First part, a, Question 2, Article 3)

The Greek influence showed itself not just in philosophy but also in science. Most importantly, Aquinas shared Aristotle's view of the universe, embracing a "geocentric" view, with the Earth at the center and the planets, Moon, and Sun going around this center, held in place by unseen crystal globes in which they are imbedded. This was obviously highly congenial to a medieval Christian such as Aquinas, who had, as it were, handed to him on a plate, the privileged place of planet Earth, the scene of God's creative works as detailed in the first book of Genesis. Philosophy, theology, and science came together in one harmonious synthesis.

THE METAPHOR UNDER THREAT

As we come toward the sixteenth century, philosophers and the great thinkers of the day began to shift how they considered the world. Whether a theory of the world as an organism, like Plato, or parts as organism-like, like Aristotle, and whether like Plato—and the Christians—that there is an actual conscious designer, or, like Aristotle, we should think of the designer more in terms of a force, a tendency, toward ends, the dominant theory was an organic metaphor or organicism. Then, a shift occurred. A new idea, the notion of the world as a machine, can be traced from the Reformation and renewed interest in ancient thinking during the

Renaissance; the machine metaphor came from the Reformation and from the Renaissance, resulting in the Scientific Revolution.

Within Christianity itself, there were factors that made the machine metaphor a logical consequence of the faith. In parallel with, and part of the rejection of the huge structure of medieval Christianity, the more biblically based thinking of Luther and Calvin—*sola scriptura*—started to loosen the hold of Greek philosophy:

> No Christian could ultimately escape the implications of the fact that Aristotle's cosmos knew no Jehovah. Christianity taught him to see it as a divine artifact, rather than as a self-contained organism. The universe was subject to God's laws; its regularities and harmonies were divinely planned, its uniformity was a result of providential design. The ultimate mystery resided in God rather than in Nature, which could thus, by successive steps, be seen not as a self-sufficient Whole, but as a divinely organized machine in which was transacted the unique drama of the Fall and Redemption. (Hall 1954, xvi–xvii)

Second, although the Renaissance brought a renewed interest in the thinking of the ancients, it was realized that not every Greek was enthused by the kind of thinking that was so important for Plato and Aristotle. In particular, there were the pre-Socratic atomists—Leucippus, Democritus, and, a little later, Epicurus. They believed that the world is made up of minute physical particles, buzzing around in the void, in empty space. In Aristotle's language, they wanted to make efficient causation all sufficient, without need of recourse to final causes. Atoms do things, and everything else follows. The best account of this position was given some centuries later by the Roman poet Lucretius, in his work *On the Nature of Things* (*De Rerum Natura*). Perhaps expectedly, he was into development, not just of individual organisms but of whole groups or species, and, as expectedly, this was all just blind chance, no purpose or end thinking:

> At that time the earth tried to create many monsters
> with weird appearance and anatomy—
> androgynous, of neither one sex nor the other but somewhere in between;
> some footless, or handless;
> many even without mouths, or without eyes and blind;
> some with their limbs stuck together all along their body,

> and thus disabled from doing harm or obtaining anything they needed.
> These and other monsters the earth created.
> But to no avail, since nature prohibited their development.
> They were unable to reach the goal of their maturity,
> to find sustenance or to copulate.
> (Sedley 2008, 150–153, *De rerum natura* V 837–848)

A person with three legs, one sticking out of the back, and with no eyes or mouth was not going to last long. But given enough time and space, even the highly improbable becomes actual:

> First, the fierce and savage lion species
> has been protected by its courage, foxes by cunning, deer by speed of flight. But as for the light-sleeping minds of dogs, with their faithful heart,
> and every kind born of the seed of beasts of burden,
> and along with them the wool-bearing flocks and the horned tribes,
> they have all been entrusted to the care of the human race. (V 862–867)

Only efficient causes here; no final causes. Eyes were not made for seeing or legs for walking. First came the eyes and legs, and then they were put to use. Denying this is to get things backwards:

> All other explanations of this type which they offer
> are back to front, due to distorted reasoning.
> For nothing has been engendered in our body in order that we might be able to use it.
> It is the fact of its being engendered that creates its use. (V 832–835)

Implausible, perhaps. But something to think about, certainly.

THE AGE OF THE MACHINE

In the Machine Age, a widely known analogy was that the world is like a timepiece, the apotheosis of machine engineering. For instance, Robert Boyle (1627–1691), the chemist, found the world to be "like a rare clock, such as may be that at Strasbourg, where all things are so skillfully contrived that the engine being once set a-moving, all things proceed according to the artificer's first design, and the motions of the little statues that at such hours perform these or those motions do not require (like those of puppets) the peculiar interposing of the artificer or any intelligent agent employed by him, but perform their functions on particular occasions by virtue of the general and primitive contrivance of the whole engine" (Boyle 1996, 12–13).

Even though the move was toward thinking of the world as being like a machine, it did not mean that people gave up being Christians or religious. This said, increasingly, people found that bringing God into the discussion was not very helpful; God was pushed out of the discussion. In the words of one of the greatest historians of the Scientific Revolution, God became a "retired engineer" (Dijksterhuis 1961, 491). In addition, during the Revolution, talk of final causes was not useful. Of course, machines do have final causes. For instance, the clock was created to tell the time. However, this part of the metaphor was dropped. In any case, argued the French philosopher René Descartes, somewhat disingenuously, one should never presume to know what end God intended: "there is an infinitude of matter in His power, the causes of which transcend my knowledge; and this reason suffices to convince me that the species of cause termed final, finds no useful employment in physical [or natural] things; for it does not appear to me that I can without temerity seek to investigate the [inscrutable] ends of God" (Descartes [1642] 1964, 111). Like many very clever people, Descartes knew when a little false modesty is in order.

REDUCTIONISM

Let us now move forward in time, from the Renaissance to the twentieth century, to test out this metaphor. We see great scientific achievements and related and subsequent technology, and a triumph of the machine metaphor: mechanism. Here, we offer one high-profile example not only to illustrate the power of the metaphor, but also to point to a very important tool or method that is an integral part of the metaphor: reductionism. In trying to understand how a machine works, usually the first thing you do is take it apart and then see what role or roles the individual parts play in the whole. You reduce the whole to its parts. (Often, "emergence" is the term used as the opposite to reduction. A new whole "emerges" as one looks at the total picture. Alternatively, reduction is spoken of as a "bottom-up" explanation, and emergence as a "top-down" explanation.)

The First World War brought home the necessity of codes and the ability to send messages that the enemy could not read or decipher. One of the most powerful machines directed to this end was the Enigma machine. Invented by a German engineer at the end of the First World War, it has a rotor mechanism that scrambles the twenty-six letters of the alphabet. The message is gobbledygook until received by another Enigma machine that can put the message back into its original readable form (Figure 1.1a). In the Second World War, Enigma was used extensively by the Germans. Building on earlier work by the Poles, the Allies learnt how to decipher the mysteries of Enigma so that they could unpack the coded messages and read their true content. Particularly in conflicts such as the Battle of the Atlantic—Allied cargo ships versus German U-boats—Enigma proved of vital worth. But how was Enigma cracked? Simply by taking it to bits and seeing how the various cogs could be understood singly as an aid to seeing how they worked together to achieve the required effects (Figure 1.1b). Reductionism!

Now take a scientific achievement: the discovery of the double helix and the uncovering of the genetic code by James Watson and Francis Crick in 1953. We see an uncanny resemblance between the nature and functioning of the Enigma machine and the molecular mechanism—molecular "machine"—by which information can be encoded, used both for building individual organisms and to transmit templates to future generations. We see also that the same reductive approach is at work. By mid-century, the theory of heredity first formulated by the monk Gregor Mendel had been accepted by all. The units of heredity, the "genes," are physical units existing in string-like entities, "chromosomes," inside cells, the building blocks of living beings. Naturally, given the molecular successes of the physical sciences, those working in the life sciences increasingly turned their attention to the ways in which the biological genes might be explained in terms of the physical molecules. The genius of Watson and Crick was to find that the "machine" at issue, the

(*a*)

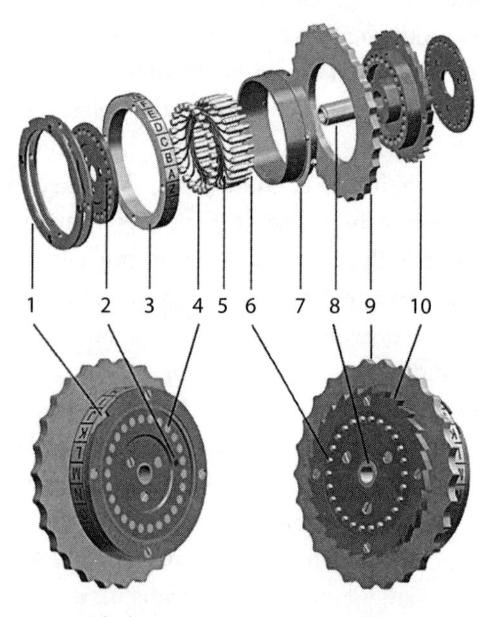

1 notched ring
2 marking dot for "A" contact
3 alphabet tyre
4 plate contacts
5 wire connections
6 pin contacts
7 spring-loaded ring adjusting lever
8 hub
9 finger wheel
10 ratchet wheel

(*b*)

FIGURE 1.1 (*a*) The Enigma machine; (*b*) The Enigma machine in parts.

molecular equivalent of the classical Mendelian gene (collectively "the genotype"), which has the information used to build the individual organism ("the phenotype") and can also transmit the information from one generation to the next, consists of two long molecules, deoxyribonucleic acid (DNA), twisted around each other to form a "double helix." The information is carried by subgroups of molecules, nucleotides, strung along the backbones of the DNA molecules. Nucleotides have four forms—adenine (A), cytosine (C), guanine (G), and thymine (T)—and it is the way that these are ordered that yields a "code" that can be (and was) deciphered, showing how the information is passed down the line (Figure 1.2a). Finding this out was a matter of isolating the nucleotides, the parts, and then seeing how they work in the system as a whole (Figure 1.2b).

Reductionism! Mechanism triumphant, right? Not quite so fast.

THE PROBLEM OF ORGANISMS

In this new way of thinking, there was a complication: the problem of organisms. Most particularly, there was the problem of final causes—what the eminent twentieth-century evolutionist Ernst Mayr spoke of as "ultimate causes" as opposed to "proximate causes." Final causes or ultimate causes, the problem is that these no longer play any role in machine-like explanations—the molecules of DNA just follow the laws of physics—but such causes are still needed to make sense of the living world. This despite the fact that in the early seventeenth century, Descartes was certain that animals are machines. The body of any animal, having "been made by the hands of God, it is incomparably better organized—and capable of movements that are much more wonderful—than any that can be devised by man, but still it is just a machine" (Descartes [1637] 1964, 41).

Robert Boyle argued that this claim is just not convincing. You can often tell the ends and, Aristotle-like, put things usefully in a final-cause context. In his *Disquisition about the Final Causes of Natural Things,* Boyle argued that supposing that "a man's eyes were made by chance, argues, that they need have no relation to a designing agent; and the use, that a man makes of them, may be either casual too, or at least may be an

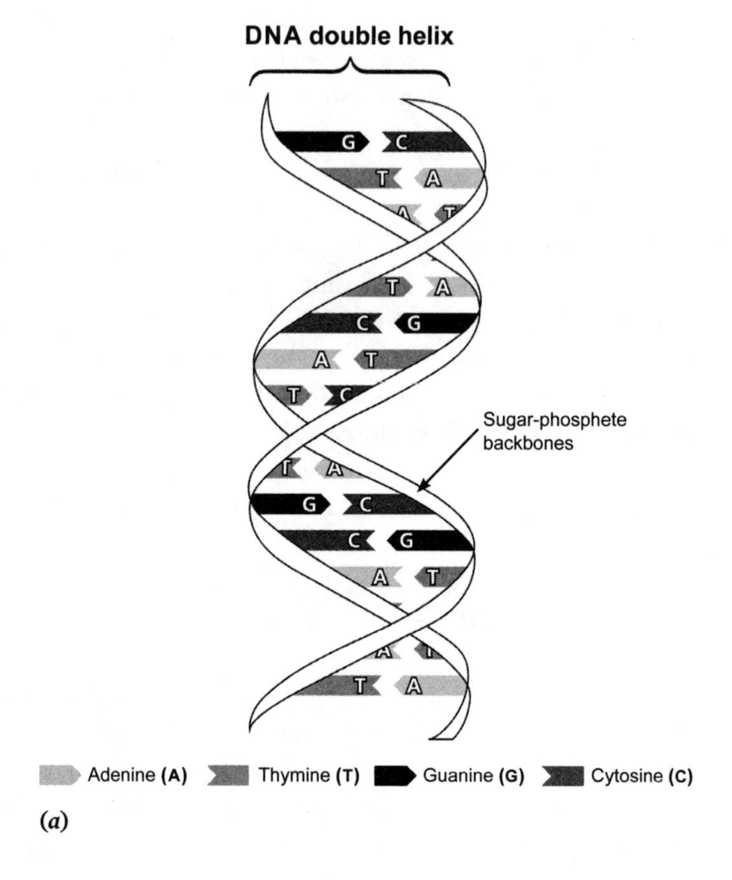

(*a*)

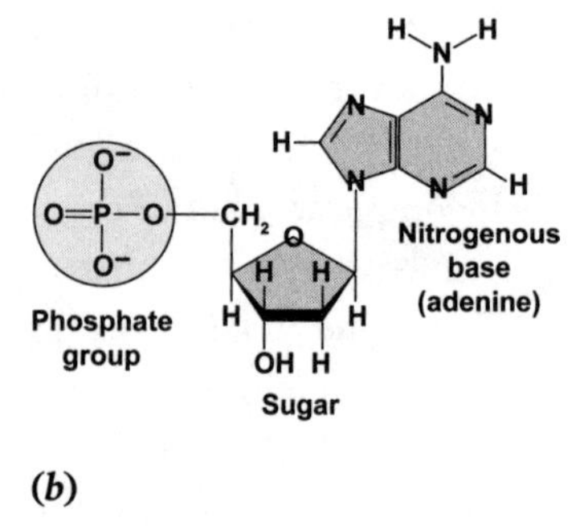

(*b*)

FIGURE 1.2 (*a*) The double helix; (*b*) A nucleotide (adenine).

effect of his knowledge, not of nature's." In Boyle's opinion, this argument is silly and counterproductive. It takes us from the chance to do science—the urge to dissect and to understand how the eye "is as exquisitely fitted to be an organ of sight, as the best artificer in the world could have framed a little engine, purposely and mainly designed for the use of seeing"—but it takes us away from the designing intelligence behind it (Boyle [1688] 1966, 397–398). Boyle found a compromise: the physical world, the material nonliving world, is a machine. Organisms are also machines, but they have an organization—a final-cause aspect—that points to a designing intelligence. God is not part of science, but science cannot do the whole job on its own and must turn to God for help. Unsurprisingly, this appeal to God in what was supposed to be a secular scientific argument was a festering sore that simply wouldn't heal. A century later, we find the great German philosopher Immanuel Kant worrying himself silly about it. In the nonliving material world, Immanuel Kant thought that Newton had done a pretty good job of things. The machine metaphor triumphed. But living things are something else. You may not like final causes, but they seem needed. And a compromise such as Boyle's is a cop-out, not a solution.

As Kant pointed out in his *Critique of Judgment,* in some very real sense, we must think of the parts of organisms as being both cause and effect. The eye, for instance, helps us to function in this world and to have offspring, which in turn brings about another eye. This fits uncomfortably with the machine metaphor. "In a watch one part is the instrument for the motion of another, but one wheel is not the efficient cause for the production of the other: one part is certainly present for the sake of the other but not because of it. Hence the producing cause of the watch and its form is not contained in the nature (of this matter), but outside of it, in a being that can act in accordance with an idea of a whole that is possible through its causality" (Kant 1790, 32). Kant goes on to speak of things using (what we shall learn is) one of our flag words, "organization." "This principle, or its definition, states: An organized product of nature is that in which everything is an end and reciprocally a means as well. Nothing in it is in vain, purposeless, or to be ascribed to a blind mechanism of nature" (33).

In the end, Kant spoke of the teleology (final-cause talk) of biology as more heuristic than descriptive of reality. Uncomfortably, he said that teleology is "therefore not a constitutive concept of the understanding or of reason, but it can still be a regulative concept for the reflecting power of judgment, for guiding research into objects of this kind and thinking over their highest ground in accordance with a remote analogy with our own causality in accordance with ends." Of course, no less than Boyle, Kant thought God was behind it all. It was just that, unlike Boyle, Kant didn't want to bring God talk anywhere near science. It didn't stop him being rather condescending about biology. "[W]e can boldly say that it would be absurd for humans even to make such an attempt or to hope that there may yet arise a Newton who could make comprehensible even the generation of a blade of grass according to natural laws that no intention has ordered; rather, we must absolutely deny this insight to human beings" (271).

EVOLUTION THROUGH NATURAL SELECTION

In the midst of debate about the world as an organism or machine and the role of God in science, we come to Charles Robert Darwin, the author of *On the Origin of Species by Means of Natural Selection or the Preservation of Favoured Races in the Struggle for Life.* Although Darwin argued for the *fact* of evolution, the slow natural unfurling of all forms from one or a few original ancestors, he was not the first to do this. Erasmus Darwin, his paternal grandfather, was one of several who were before him:

> From this account of reproduction it appears, that all animals have a similar origin, viz. from a single living filament; and that the difference of their forms and qualities has arisen only from the different irritabilities and sensibilities, or voluntarities, or associabilities, of this original living filament; and perhaps in some degree from the different forms of the particles of the fluids, by which it has been at first stimulated into activity. (E. Darwin 1794–1796, *Zoonomia* I 6)

Charles Darwin's major contribution in this debate was to come up with a plausible *cause*—a force akin in some way to Newton's force of gravitational attraction—to explain this fact of evolution. To this end, he seriously considered what the philosophers of scientific methodology of his day were saying about Newton's force of gravitational attraction. Above all, he learned that Newton had provided the paradigmatic example of a *vera causa,* a "true cause," and that such a cause must, Newton-like, be embedded in a kind of axiom system with the causes at the top and the effects at the bottom—Newton's law of gravitational attraction at the top, Kepler's laws of planetary motion at the bottom. But what does that all mean exactly? There was some ambiguity here. The astronomer John Frederick William Herschel (1830), somewhat inclined to empiricism—we work *from* experience of the natural world—argued that the mark of a true cause was analogy—we reason from something we know to something we don't know. How do we know a force is pulling the Moon toward the Earth? Because we have felt such a force when we spin around a stone tied to the end of a piece of string. William Whewell (1840), historian and philosopher of science, inclined to rationalism—we work *to* experience of the natural world—argued that we don't need such direct evidence. It is enough that the supposed true cause can explain many different items of evidence—as the killer is identified from the clues, the bloodstains, the knife, the broken alibi.

Darwin took note of these factors and built them all into the *Origin* (Ruse 1975). First, speaking to Herschel's dictates, he started with the analogy of farmyard and fanciers' breeding and suggested it as an analogy for what happens in nature. Then, having supposed that new variations—the building blocks of evolution—appear constantly in populations, he argued—in at least an informal deductive axiomatic sort of way—for what the Anglican clergyman and political theorist Thomas Robert Malthus ([1826] 1914) called a "struggle for existence." Premise 1: Organisms reproduce at a geometric rate. Premise 2: Food and space can be increased only arithmetically. Premise 3: two times two times two is bigger than two plus two plus two. Conclusion: There will be a struggle for life and even more for reproduction. Hence, next generations will be from just a limited subgroup of the parent generations, and that will mean a kind of natural selection, as success in the struggle is on average a result of having more helpful variations than rivals: "can we doubt (remembering that many more individuals are born than can possibly survive) that individuals having any advantage, however slight, over others, would have the best chance of surviving and of procreating their kind? On the other hand, we may feel sure that any variation in the least degree injurious would be rigidly destroyed. This preservation of favourable variations and the rejection of injurious variations, I call Natural Selection" (Darwin 1859, 80–81).

From natural selection, Darwin went on to argue that the result of evolution is the famous tree of life. "The affinities of all the beings of the same class have sometimes been represented by a great tree. I believe this simile largely speaks the truth. The green and budding twigs may represent existing species; and those produced during each former year may represent the long succession of extinct species" (129). This doesn't just happen by chance. Drawing on the insight of the eighteenth-century Scottish political economist Adam Smith that for efficient working you need a division of labor, Darwin argued that speciation comes about as organisms look for different noncompeting niches and diverge away from each other. All selection driven.

For the second half of the *Origin,* Darwin set about showing how evolution through natural selection could function as a kind of Whewellian *vera causa*—exhibiting what Whewell called a "consilience of inductions"—explaining across the range of the life sciences (Figure 1.3). Social behavior is a function of selection working on the whole nest rather than just on individuals in the group. The fossil record, paleontology, is easy. Darwin (1859, 435) spoke comfortably of "the ancient progenitor, the archetype as it may be called, of all mammals." Geographical distribution also follows. Darwin took special note that the organisms of the Galapagos look like those of South America and those of the Canaries like those of Africa. Systematics, morphology, and anatomy fell into place—as did embryology. Why are often-remarked similarities of the embryos of organisms so different as adults? Because the embryos are protected from selection in their early stages and only feel the effects of selection as they grow toward adulthood.

Darwin was ready for his final, triumphant conclusion:

> [F]rom the war of nature, from famine and death, the most exalted object which we are capable of conceiving, namely, the production of the higher animals, directly follows. There is grandeur in this view of life, with its several powers, having been originally breathed into a few forms or into one;

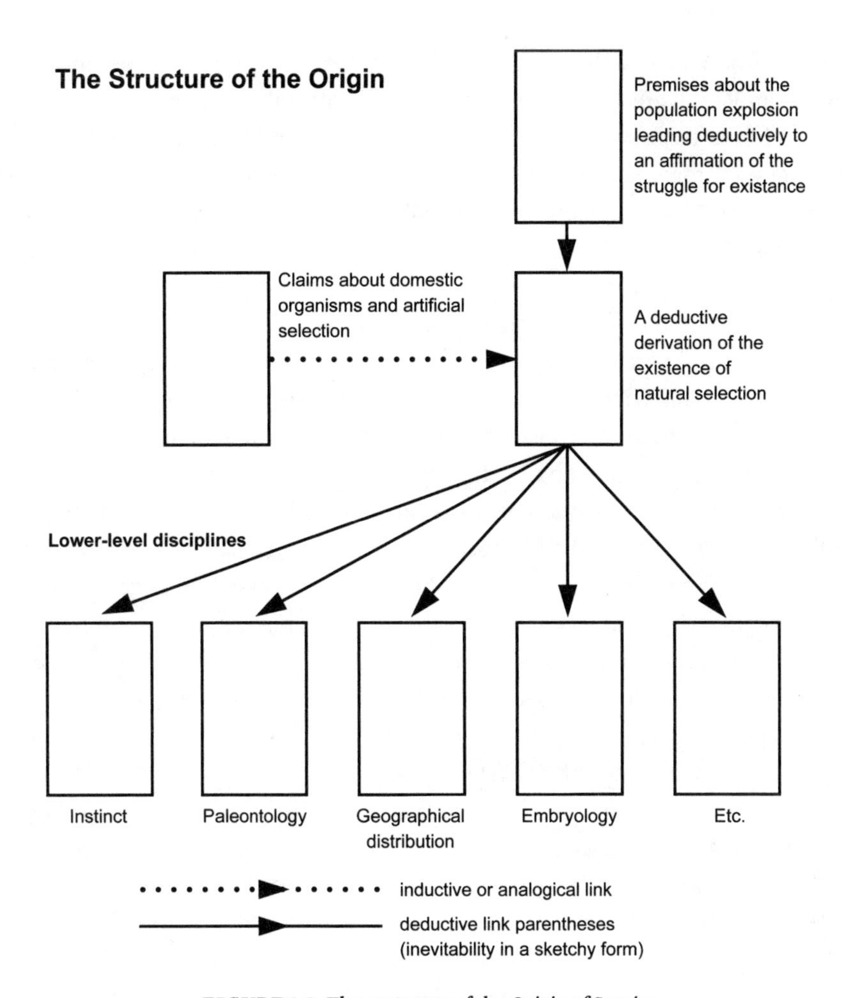

FIGURE 1.3 The structure of the *Origin of Species*.

> and that, whilst this planet has gone cycling on according to the fixed law of gravity, from so simple a beginning endless forms most beautiful and most wonderful have been, and are being, evolved. (489–490)

PUTTING DARWIN IN CONTEXT

Let us put Darwin's theory into the context of ideas and terms introduced already. Darwin was a mechanist. He flaunted his commitment to a strictly law-bound explanation of everything. Opposite his title page, he quoted William Whewell. "But with regard to the material world, we can at least go so far as this—we can perceive that events are brought about not by insulated interpositions of Divine power, exerted in each particular case, but by the establishment of general laws." This was a little naughty because Whewell (1837)—no evolutionist he—played Robert Boyle's trick of taking organic origins out of the realm of science—"science says nothing, but she points upwards."

Later in life—ten or so years after he wrote the *Origin*—Darwin embraced what his friend and supporter Thomas Henry Huxley called "agnosticism," an inability to say if there is or is not a deity. However, when formulating and publishing his theory, Darwin embraced a form of deism, in a statement he left unchanged through the six (very much changed) editions of the *Origin* (the final sixth edition was in 1872). "To my mind it accords better with what we know of the laws impressed on matter by the Creator, that the production and extinction of the past and present inhabitants of the world should have been due to secondary causes, like those determining the birth and death of the individual" (Darwin 1859, 488).

What of reductionism? In respects, one might say that Darwin was the very opposite of a reductionist, being rather a holist, where this is understood as seeing things in the context of the entire system—following Aristotle in arguing that the "whole is greater than the sum of its parts" (Aristotle 980a *Metaphysics*). Emergence! The tree of life seems to bear this out. But pause for a moment and ask how the tree comes about. As we have seen, it is through branching, and this is brought on by the division of labor, with organisms specializing and then having to adjust to others. The division of labor is set entirely in the eighteenth-century world of the Scottish businessman, whom Adam Smith ([1776] 1937) is trying to describe and understand, and the first premise is that no one does anything for anyone else without hope of return. In fact, no one does anything at all without hope of return. "This division of labour, from which so many advantages are derived, is not originally the effect of any human wisdom, which foresees and intends that general opulence to which it gives occasion. It is the necessary, though very slow and gradual consequence of a certain propensity in human nature which has in view no such extensive utility; the propensity to truck, barter, and exchange one thing for another" (Smith 1776, 2: 1). Notoriously, Smith added: "It is not from the benevolence of the butcher, the brewer, or the baker that we expect our dinner, but from their regard to their own interest."

Charles Darwin bought entirely and completely into this. Why wouldn't he? He was the grandson (on his mother's side) of one of the most successful businessmen of the age—Josiah Wedgwood—and it was the family creed, both Darwins and Wedgwoods. Darwin always saw natural selection working to the good of the individual not the group. "Hence, as more individuals are produced than can possibly survive, there must in every case be a struggle for existence, either one individual with another of the same species, or with the individuals of distinct species, or with the physical conditions of life" (1859, 64). In this respect, he was no holist but rather an eager reductionist. Darwin did have to think hard about possible exceptions. In the *Origin,* he wrestled with the sterile-insect problem. In the nests of the hymenoptera—ants, bees, and wasps—one finds workers, sterile females, devoting their lives to the good of the group. How can this be reconciled with a theory that stresses above all the importance of reproduction? Darwin did not have a proper theory of heredity, so he could not really tease out what was going on. But he spotted that it had to be something to do with relatives. Somehow, when a relative reproduces, that is good for you—because somehow you are biologically alike. In our terminology today, we would say that you share some of the same genes. Darwin's solution was to treat the nest or hive as an individual, a kind of superorganism, whereas in organisms, the parts work for the whole—liver, kidneys, heart, all joined in the same mission. Simply, if they don't make it, you won't make it. So, he could keep up the individual selection stance. It was all a matter of what constituted an individual. Today, this form of selection is known as "kin selection."

Reductionism rules, OK! What about final cause, teleology? One thing that you can say is that Darwin's theory is as final-cause—ultimate-cause—impregnated as anything in Aristotle. The question is: in what sense? As the distinguished evolutionist Ernst Mayr noted any Darwinian analysis is going to be bound up with natural selection:

> [Final or ultimate causes are causes] that have a history and that have been incorporated into the system through many thousands of generations of natural selection. It is evident that the functional biologist would be concerned with analysis of the proximate causes, while the evolutionary

> biologist would be concerned with analysis of the ultimate causes. This is the case with almost any biological phenomenon we might want to study. There is always a proximate set of causes and an ultimate set of causes; both have to be explained and interpreted for a complete understanding of the given phenomenon. (Mayr 1961, 1503)

In the discussion of embryology in the *Origin,* Darwin asks why in some species—as opposed to others such as butterflies—the young closely resemble the adults:

> With respect to the *final cause* of the young in these cases not undergoing any metamorphosis, or closely resembling their parents from their earliest age, we can see that this would result from the two following contingencies; firstly, from the young, during a course of modification carried on for many generations, having to provide for their own wants at a very early stage of development, and secondly, from their following exactly the same habits of life with their parents; for in this case, it would be indispensable for the existence of the species, that the child should be modified at a very early age in the same manner with its parents, in accordance with their similar habits. (Darwin 1859, 447–448, our italics)

Final causes. For a selection-based explanation of final (ultimate) cause, Mayr prefers to use the term "teleonomy" rather than "teleology." He feels that this change marks the sense that now we are working under the machine metaphor rather than the organism metaphor. Most scholars today, however, are comfortable with continued use of "teleology," marking the fact that we still have explanations in terms of forward looking—what we expect to happen. Either way, Darwin did not deny teleology (teleonomy). He talked happily of final causes, meaning that it makes good sense to ask the purpose of features of embryological development, in this case of the earliest stages looking like the later stages. The question is why this lack of change is not a product of an external designer or of internal forces. Nor is the teleology purely heuristic. In Darwin's efficient-cause driven world, organic features are (as Mayr stressed) products of natural selection. Uniformity in development helped the possessor in the past, so we assume it will help the possessor in the future. We could be wrong, but that is how we think, and how we will continue to think unless and until changed circumstances introduce new selective pressures.

We have dated the end of the Scientific Revolution with the appearance of Newton's *Principia.* It would be idiosyncratic not to do so. We have seen, however, that with respect to the problem of organisms and their need of final-cause explanations, the Revolution was in this sense incomplete. Darwin is important because, in this respect, he completed the Scientific Revolution (Ruse 2017, 2021a). Final causes can be subsumed under the machine metaphor. The world is to be seen mechanistically. And note what this means. You can have relative value but not absolute value. At least, you cannot get it from the science. You can at once see the implications. In the *Origin,* to avoid controversy, Darwin made only brief reference to our species, *Homo sapiens,* and that only that he not be accused of cowardice. "In the distant future I see open fields for far more important researches. Psychology will be based on a new foundation, that of the necessary acquirement of each mental power and capacity by gradation. Light will be thrown on the origin of man and his history" (488). Later, in his *Descent of Man* (1871), Darwin turned his full attention to our species, arguing that we too are the product of natural forces, although in this later book, Darwin gave a great causal role to the secondary mechanism of sexual selection. The implication, however, is as before. No absolute values. Thanks to their intelligence and other distinctive attributes, you may value humans above all others. We certainly do! But you are not going to get it out of Darwinian evolutionary theory. Nor should you expect to (Ruse 2021a).

End of story? Well, not quite! The organic model fell out of favor, but there were always those who thought of it favorably. Not surprisingly, these were often people who identified with Plato, for instance the group of seventeenth-century English philosophers who, noting both the influence and their location, were known as the Cambridge Platonists. The real renaissance for the organic model came, however, with the German Romantics, often in science designated as advocates for *Naturphilosophie*—a stance that called for a replacement of "the concept of mechanism" and a renewal of the organic metaphor, "elevating it to the chief principle for interpreting nature" (Richards 2002, xvii). This is our topic for Chapter 2.

Chapter Two

ORGANICISM STRIKES BACK

> As the body displays tears, mucus, and earwax, and also in places lymph from pustules on the face, so the Earth displays amber and bitumen; as the bladder pours out urine, so the mountains pour out rivers; as the body produces excrement of sulphurous odor and farts which can even be set on fire, so the Earth produces sulphur, subterranean fires, thunder, and lightning; and as blood is generated in the veins of an animate being, and with it sweat, which is thrust outside the body, so in the veins of the Earth are generated metals and fossils, and rainy vapor.
>
> —KEPLER [1619] 1977, 363–364

GERMAN ROMANTICISM

The German astronomer Johannes Kepler is rightfully considered one of the heroes of the Scientific Revolution, for it was he who broke from 2,000 years of thinking that the heavens must go on paths of circular perfection. He showed that, in fact, the planets trace ellipses with the Sun at one focus. He was a leader in convincing of the power of the machine metaphor. Yet, as the quotation above shows only too clearly, things are never quite that simple. Kepler was obsessed with mathematics, worshipping the memory of the great Pythagoras, who considered the Sun all-important, a view taken up by his disciple Plato and incorporated into his metaphysical world picture. Little wonder then that Kepler, for all he made major moves forward under the machine metaphor, happily also embraced the organic metaphor. Even less wonder, Kepler acknowledged his debt to Pythagorean thinking, as shown by Plato. "The view that there is some soul of the whole universe, directing the motions of the stars, the generation of the elements, the conservation of living creatures and plants, and finally the mutual sympathy of things above and below, is defended from the Pythagorean beliefs by Timaeus of Locri in Plato" (Kepler [1619] 1977, 358–359).

Do not think that Kepler's organicism was just a remnant from which a clean break would be made as time rolled on. To the contrary, to use a self-reflecting metaphor, organicism proved a hardy plant, especially—to take up from the end of the last chapter—as we come to the end of the eighteenth century and encounter the rise of Romanticism. Notably, the great poet Johann Wolfgang von Goethe believed in the holistic nature of living beings. Stressing the supposed ways in which the vertebrate skull is made from modified identical parts, originally vertebrae, he is a good example of a German Romantic enthusiast of the organic model. In the plant world, he argued for shared development of plants, from a basic ground plan or archetype (Figure 2.1)—the *Urpflantz* (Goethe [1790] 1946).

Goethe was important, as were others such as the anatomist Lorenz Oken. However, it was the philosopher Friedrich Schelling who truly ran with the Platonic vision of the whole world as an organism:

> Ascending up to thought's youthful force,
> Whereby nature continually renews its course,
> There is one power, one pulse, one life
> A continual exchange of resistance and strife.
> (Schelling 1962–1967, *Epikurisch Glaubensbekenntniss Heinz Widerporstens*, cited by Richards 2002, 310)

Schelling wanted to break down the distinction between the objective—the world out there—and the subjective—the world in here. In Platonism, particularly in the theory of Forms or Ideas, Schelling saw how this insight could be explicated, since both objective reality and personal subjectivity have their being in the mind of the Demiurge. "The key to the explanation of the entirety of the Platonic philosophy is

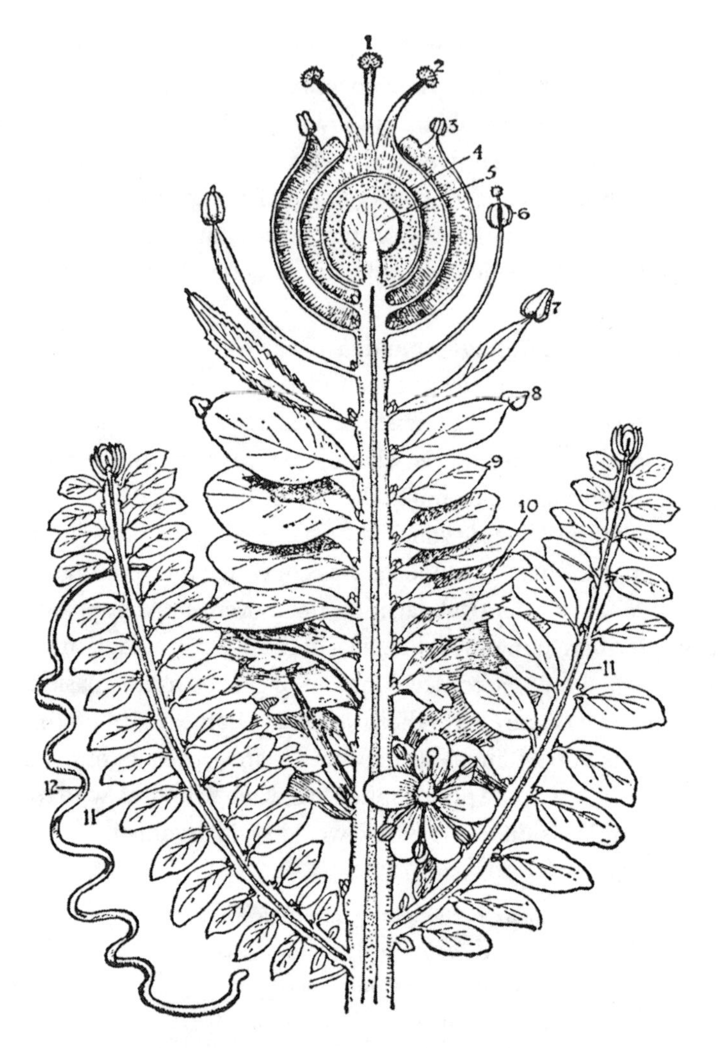

FIGURE 2.1 Goethe's *Urpflanz*. *Credit:* Reproduced from Johann Wolfgang Goethe, *Versuch die Metamorphose der Pflanzen zu erklären* (Gotha: Ettingersche Buchhandlung, 1790).

noticing that Plato *everywhere carries the subjective over to the objective*" (Schelling [1833–1834] 1994, 212, his italics). An implication of this was that Schelling rejected the Kantian judgment that the end-directedness of organisms decrees the second-rate nature of biological understanding—that the only good explanations are mechanistic explanations. This would not do, could not do. If one side of human awareness and understanding is the subjective, then at some level, this must be reflected in the other side, the objective. If final-cause thinking is needed in our science, as it is, then it must exist in some real sense out in the world.

In other words, the physical world must be essentially organic, with final cause an essential part of it. "Even in mere organized matter there is *life,* but a life of a more restricted kind. This idea is so old, and has hitherto persisted so constantly in the most varied forms, right up to the present day—(already in the most ancient times it was believed that the whole world was pervaded by an animating principle, called the world-soul, and the later period of Leibniz gave every plant its soul)—that one may very well surmise from the beginning that there must be some reason latent in the human mind itself for this natural belief" (Schelling [1797] 1988, 35). We have just had a reference to organization, and this makes the case complete. The world is something that produces itself, has its developing powers inside, as an unfurling organism is driven by forces within rather than without. One goes from the simple to the complex, from the undifferentiated to the highly differentiated. "Nature should be Mind made visible, Mind the invisible nature. Here then, in the absolute identity of Mind *in us* and Nature *outside us,* the problem of the possibility of a Nature external to us must be resolved. The final goal of our further research is, therefore, this idea of Nature; if we succeed in attaining this, we can also be certain to have dealt satisfactorily with that Problem" (42).

HUMANS

What about us? Where do we humans fit into the picture? Very differently from the mechanist, who has drained the picture of the world of absolute value, the status of humans is the golden thread running through the organicist world picture. The organism is conceived and born, it grows, and finally becomes the mature adult. From acorn to oak, from caterpillar to butterfly, from helpless infant to philosopher king. The whole picture, the whole metaphor, is one of growth, of improvement, of going from value-empty to value-full. The special status of humans, at the head of the living world, comes with the territory. According to Plato, "God gave the sovereign part of the human soul to be the divinity of each one, being that part which, as we say, dwells at the top of the body, inasmuch as we are a plant not of an earthly but of a heavenly growth, raises us from earth to our kindred who are in heaven." Aristotle: "after the birth of animals, plants exist for their sake, and that the other animals exist for the sake of man . . . Now if nature makes nothing incomplete, and nothing in vain, the inference must be that she has made all animals for the sake of man" (Barnes 1984, 1256b15–22). The Christian philosophers likewise bought into the superiority of humans: "God created mankind in his own image, in the image of God he created them; male and female he created them" (Gen. 1: 27).

Interpreted in a Christian context or not, this kind of thinking is the very essence of Romanticism. It is a fundamental part of Schelling's philosophy. This value-impregnated view of evolution has but one designated end point. "It is One force, One interplay and weaving, One drive and impulsion to ever higher life" (Schelling, *Proteus of Nature,* 1800, in Morgan 1990, 35). This tradition continued down through the nineteenth century. After the *Origin,* the best-known, certainly the most enthusiastic, proponent of this kind of thinking was the evolutionist Ernst Haeckel (Richards 2008). Stressing the organicist nature of his thinking with his "biogenetic law"—ontogeny (development of the individual organism) recapitulates phylogeny (evolution of the line)—he was ever into drawing trees of life, with humans at the top. More accurately, with European humans at the top (Haeckel 1896).

HERBERT SPENCER

Romantic organicism was incredibly infectious, especially as evolutionary thinking became more and more popular and acceptable. This last point is no great surprise, for there is reason to think that virtually all the major Romantics, starting with Goethe, were or became evolutionists. With the organic emphasis on development and growth, evolutionary ideas come with the territory. The English contemporary of Darwin, Herbert Spencer, is an exemplar. Not a good start, you might think, because Spencer has a reputation as the epitome of the selfishness approach to life, a founder of what has become to be known as "Social Darwinism," a particularly stringent form of the *laissez-faire* philosophy of society. Margaret Thatcher, squared and cubed (O'Connell and Ruse 2021). There are grounds for this reputation. Those who tried to ameliorate the condition of the poor were objects of scorn to Spencer: "Blind to the fact that under the natural order of things, society is constantly excreting its unhealthy, imbecile, slow, vacillating, faithless members, these unthinking, though well-meaning, men advocate an interference which not only stops the purifying process but even increases the vitiation" (Spencer 1851, 323–324).

This was written before Spencer became an evolutionist. Even then, to flag Spencer as the proponent of a particularly vile social philosophy is a mistake. Like Margaret Thatcher, his emphasis was always far more on the success of the successful rather than the failure of the failed. The underlying belief was that striking down barriers to advancement, which inevitably favored the rich and established, would level the playing fields, so that the talented and energetic could rise and benefit. Already, the underlying belief was in progress. When Spencer became an evolutionist, this was interpreted increasingly strongly in terms of German Romantic holism. Spencer argued that societies are organisms, with parts, and saw this throughout the living world (Spencer 1860). Never that keen on selection—for all that he grasped it independently of Darwin—Spencer was ever a follower of the early-nineteenth-century French evolutionist, Jean Baptiste de Lamarck (Spencer 1852a, [1852b] 1868). Best known for his embrace of the inheritance of acquired characteristics—the parent giraffe stretches up for leaves on the tree, and the offspring giraffe is born with a longer neck—Lamarck drew heavily on the *Naturphilosophen,* proposing some kind of internal force, *Le pouvoir de la vie,* leading to ever-greater complexity (Lamarck 1815–1822). For Spencer, therefore, although the struggle for existence leads to change and progress, its action is not primarily through selection, as it is for Darwin, but through the struggle leading to effort and inheritable change. Apparently, there is only so much vital fluid available, and it goes to making babies or brains—as we move upwards, fertility drops away—herrings to humans—and the struggle ceases and all work for the common good. Apparently also fueling change are outside disruptions that take groups from stability and stasis and cause disruption and consequent change and upwards movement, from the homogeneous (all the same) to the heterogeneous (differences). This Spencer (1862) called his theory of "dynamic equilibrium."

Spencer was much impressed by recent advances in thermodynamics and felt that he was in a sense simply translating physics into the biological world. Notwithstanding, all of this is a highly organicist vision—as it was for Schelling—a debt Spencer acknowledged: "the acquaintance which I accidently made with Coleridge's essay on the Idea of Life, in which he set forth, as though it were his own, the notion of Schelling, that Life is the tendency to individuation, had a considerable effect" (Duncan 1908, 541). Expectedly, Spencer finds value in the very process. As evolution moves upwards, things are improved. "Ethics has for its subject-matter, that form which universal conduct assumes during the last stages of its evolution" (Spencer 1879, 21). Continuing: "And there has followed the corollary that conduct gains ethical sanction in proportion as the activities, becoming less and less militant and more and more industrial, are such as do not necessitate mutual injury or hindrance, but consist with, and are furthered by, co-operation and mutual aid."

"Value in the very process." We are no longer in the world of the mechanist. Most obviously and importantly, we are in the world of progress, with the end point humankind. "This law of organic progress is the law of all progress. Whether it be in the development of the Earth, in the development of Life upon its surface, in the development of Society, of Government, of Manufactures, of Commerce, of Language, Literature, Science, Art, this same evolution of the simple into the complex, through successive differentiations, holds throughout" (Spencer [1857] 1868, 245). Spencer explains that the English language is more complex and hence above all others. Far be it for the English-born authors of this work to disagree!

VITALISM

Schelling's influence was felt in the New World as well. Swiss-born Louis Agassiz, student of Schelling, moved to Harvard, the home of New England transcendentalism, a philosophical movement with deep roots in German Romanticism. As it happens, Agassiz was more conservative than most and could never bring himself to accept evolution. This did not stop his students, including his own son Alexander, from becoming organismic evolutionists (Lurie 1960; Winsor 1991). This was just the start of things. In the next century, the Harvard faculty was enriched by the arrival of the English logician Alfred North Whitehead—who, with Bertrand Russell, was deservedly famous for the attempt (in their magnum opus *Principia Mathematica*) to show that mathematics follows deductively from the laws of logic. In the 1920s, moving into metaphysics, Whitehead gave a series of lectures, published as *Science and the Modern World*. Openly declaring himself an organicist, he called for "the abandonment of the traditional scientific materialism, and the substitution of an alternative doctrine of organism" (Whitehead 1926, 99). Continuing: "Nature exhibits itself as exemplifying a philosophy of the evolution of organisms subject to determinate conditions" (115). There is no need to search far for the influences. "Whitehead's critique of scientific materialism and his philosophy of organism can be interpreted as efforts to develop Schelling's ideas more rigorously in the light of recent physics. For Whitehead, as for Schelling, nature is 'unconscious mind'" (Gare 2002, 36). Continuing: "Schelling's evolutionary cosmology in which nature is seen as self-organizing . . . formed the core of Herbert Spencer's evolutionary theory of nature, which then had a major influence on both Bergson and Whitehead."

Bergson? The reference is to Henri Bergson, the French philosopher, author of *L'évolution créatrice*, published in 1907 (English translation 1911), champion of the neo-Aristotelian life force, the *élan vital*—hence, better known as a "vitalist" rather than the more comprehensive "organicist." The philosophy is the same and is derivative: deeply Aristotelian, including the importance of final cause. "The 'vital principle' may indeed not explain much, but it is at least a sort of label affixed to our ignorance, so as to remind us of this occasionally, while mechanism invites us to ignore that ignorance" (Bergson 1911, 42). Expectedly, vitalism speaks to "internal finality." And, linking back, one is not surprised to find more immediate and more familiar influences. Jean-Gaspard-Félix Laché Ravaisson-Mollien, arguably the most influential French philosopher of the second half of the nineteenth century, was a student of Schelling and a teacher of Bergson. Gilding the lily, while being educated at the École Normale,

Bergson became an enthusiastic Spencerian. With predictable conclusions: "not only does consciousness appear as the motive principle of evolution, but also, among conscious beings themselves, man comes to occupy a privileged place. Between him and the animals the difference is no longer one of degree, but of kind" (Bergson 1911, 34). More than this even: "in the last analysis, man might be considered the reason for the existence of the entire organization of life on our planet" (35).

THE COMING OF GENETICS

Time for a scientific interlude. Darwin gave us the theory of evolution through natural selection. What he did not give us was an adequate theory of heredity. As everyone now knows, in the 1860s, while Darwin was looking frenetically for a solution (Darwin 1868)—and while his critics were having a field day—in part of the Austro-Hungarian Empire, a plant-breeding monk, the earlier-mentioned Gregor Mendel, was uncovering what we now believe are the true principles of heredity (Bowler 1989). His two laws of transmission showed how the units of heredity, what he called "factors" and we call "genes," can be passed on from generation to generation. However, Mendel's work went unappreciated—he himself, on reading an early translation of the *Origin,* was more concerned whether he, a Catholic priest, could accept evolution than arguing that he had found the solution to Darwin's problems. It was not until the beginning of the twentieth century, by which time people knew a lot more about the nature of the cell, that three people independently discovered Mendel's work, giving him credit for what he had done.

Nothing in this life goes smoothly, and at first, the champions of selection—the "biometricians"—and the champions of Mendel—the "hereditarians," who came to be known as the "geneticists"—thought they had rival theories. Only one could be right—Darwin or Mendel (Provine 1971). It was not until the 1920s that people started to realize that the two positions are not conflicting rivals but rather complementary parts of the whole picture. Crucially important was the American Thomas Hunt Morgan and his students working at Columbia University in New York City. Thanks to him, people got a better physical grasp of the hitherto theoretical gene (Allen 1978). As we saw in Chapter 1, it was seen as something physical on the string-like entities in the center (nucleus) of the cell, the chromosomes. And the number and modes of action were precisely those postulated and needed by Mendelian genetics.

At this point, gifted mathematicians—theoretical population geneticists—showed how evolution is all a matter of selection (and some other factors, such as mutation, the spontaneous generation of new variations) working on populations of organisms, happy possessors of Mendelian genes. In Britain, the two key figures were Ronald A. Fisher (1930) and J. B. S. Haldane (1932); in the United States, it was Sewall Wright (1931, 1932). The mathematicians gave us the theoretical populational genetical theory of evolution. Then, the empiricists—the naturalists and experimentalists—moved in and, by around 1940 or so, were well on the way to formulating what came to be known in Britain as Neo-Darwinism and in the United States as the synthetic theory of evolution, being a synthesis of the ideas of Darwin and Mendel. In Britain, key figures included E. B. Ford and his students, founding what came to be known as the school of "ecological genetics" (e.g., Fisher & Ford [1947] 1974; Kettlewell 1955; Cain 1954; Sheppard 1958; Ford 1964). In the United States, all important were the Russian-born geneticist Theodosius Dobzhansky (author of *Genetics and the Origin of Species,* 1937) and, following him, the already introduced German-born ornithologist Ernst Mayr (*Systematics and the Origin of Species,* 1942), the paleontologist George Gaylord Simpson (*Tempo and Mode in Evolution,* 1944), and, a little later, the botanist G. Ledyard Stebbins (*Variation and Evolution in Plants,* 1950). The year 1959 was the hundredth anniversary of the *Origin,* and there were general celebrations marking what was taken as the arrival of a mature theory of evolution (Smocovitis 1999).

ORGANICISM: PERHAPS BLOODY, CERTAINLY UNBOWED

In pure science, was mechanism really triumphant? You would think so if you read the speeches given at the major Darwin-honoring gathering at the University of Chicago. (Honorary degrees all around to celebrate!)

In the language of Thomas Kuhn (1962), the revolution is now finally over, and evolutionary theorizing has its "paradigm." Well, not so fast! Although far from a majority position, organicism moved almost comfortably into and down the twentieth century. Negatively, mechanism was anathematized. Just at the time when the experimentalists were taking over from the mathematicians, we get:

> Soon after the middle of last century it thus became a general belief among most biologists and numerous popular writers that life must ultimately be regarded as no more than a complicated physico-chemical process. In the writings of [Thomas Henry] Huxley, for instance, we find this belief very clearly formulated.

This was a triumph of optimism over reality.

> When, however, we attempt to form any detailed conception of what sort of physico-chemical process could, on the prevailing mechanistic conception of physics and chemistry, correspond with the characteristic features of life, the attempt breaks down completely. We can form no conception on these lines of how it is that a living organism, presuming it, as we must on the mechanistic theory, to be an extremely complex and delicately adjusted piece of molecular machinery, maintains and adjusts its characteristic form and activities in face of a varying environment, and reproduces them indefinitely often. This is so evident that there is no need to enlarge upon it. (J. S. Haldane 1935, 16)

These are the words of J. S. Haldane, physiologist father of J. B. S. Haldane. One suspects something deeply Oedipal going on here!

Positively, holism was extolled. Its greatest theoretical champion, a man whose chief claim to fame is that first he fought successfully against the British and then successfully with the British was the military general and sometime prime minister of South Africa, Jan Smuts. Trained as a philosopher, he wrote of the "fundamental holistic characters as a unity of parts which is so close and intense as to be more than the sum of its parts"; from this, it follows that "the whole and the parts, therefore reciprocally influence and determine each other, and appear more or less to merge their individual characters: the whole is in the parts and the parts are in the whole, and this synthesis of whole and parts is reflected in the holistic character of the functions of the parts as well as of the whole" (Smuts 1926, 88). This is the crunch point. Is the aim of the best science to break down and then to explain? To be a reductionist? Or is it to be an emergentist? To take the whole as fundamental, and one is building up to this. To be a holist? Top-down rather than bottom-up explanation?

This is not just an epistemological question. There are ethical issues here too. If it is all a question of breaking down, then trees are just molecules, and you can treat them as you will. Who cares about overall effects? If trees are more than molecules and have personal standing, then might they have rights? Can we just destroy the environment for our own pleasure? If animals are just molecules, bring on the steaks. If animals have standing as wholes, then become a vegetarian and perhaps even a vegan. (For now, hold these thoughts. We will return to them in later chapters.) Against the background of this kind of thinking—anti-mechanism, pro-holism—more than one eminent evolutionary biologist felt free to dip not just their toes into the invitingly warm water offered by Spencer and Bergson. Let us go straight to the top.

SEWALL WRIGHT

In the early twentieth century at Harvard, many of the biologists were Spencer enthusiasts. They believed strongly in group processes, including selection, and they saw life as moving upwards, from equilibrium point to equilibrium point, including, especially including—as for Spencer—humankind. One graduate of Harvard—his father was on the faculty—who bought into this line completely was the hugely important and influential already-encountered American population geneticist, Sewall Wright (Provine 1986). Deservedly, Wright is considered a major figure in the arrival of modern evolution theory. Yet, he was ever a bit half-hearted about natural selection, following Spencer in being unsure if it could produce the needed change for evolution. At a National Research Council conference in 1955 to discuss biological concepts, participants were asked to say which concepts they thought most significant. Wright tagged "organization," "replication," "variation," "evolution" (the fact not the cause), and "hierarchy." He ignored "natural selection" (Ruse 2021a).

What alternative did Wright offer? His "shifting balance" theory of evolution supposed that species are in equilibrium. Something untoward happens—a new predator, unfavorable climatic conditions, disease—and the species fragments into small groups. In such cases, Wright showed that cumulative effects of undirected breeding can outweigh selection. This he called "genetic drift," and it is here that new innovations come into being—not through selection, as for the orthodox Darwinian, but rather by chance as it were. This point was incidentally seized on by paleontologists for whom, dealing as they were with long-dead organisms, natural selection never seemed that pressing. This lay behind the much-discussed paleontological hypothesis of "punctuated equilibria" (Eldredge and Gould 1972). For Wright, who (focusing on living organisms) had no objection to selection in principle, crisis over or conquered, the small groups come back into contact, and there is a kind of intraspecific group selection as more successful groups (thanks to their randomly obtained innovations) push out others and become the species norm. We are taken back to equilibrium. Ring a bell? Consider the comment Wright makes at the end of the paper introducing this theory: "The present discussion has dealt with the problem of evolution as one depending wholly on mechanism and chance. In recent years, there has been some tendency to revert to more or less mystical conceptions revolving about such phrases as 'emergent evolution' and 'creative evolution'" (Wright 1931, 155). He rushes to say that he thinks such ideas have no place in science, but he "must confess to a certain sympathy with such viewpoints philosophically." He certainly did, for his theory is Herbert Spencer in mathematical form (Ruse 2004)!

JULIAN HUXLEY

Meanwhile, back in the country of Charles Darwin's birth . . . In 1942, Julian Huxley, brother of the novelist Aldous Huxley, grandson of Darwin's "bulldog," Thomas Henry Huxley, published his book, *Evolution: The Modern Synthesis,* introducing a unified idea to the world. Trumpets blowing loudly, the "modern synthesis," supposedly the *Origin* updated, brought together ideas across many areas of biology, bridging the gaps between genetics, natural science, and evolutionary science, among others. Yet, to the hardline Darwinian, there was a worm in the bud—an organicist worm in the mechanistic bud. From the beginning of his career, a biologist following in his grandfather's footsteps, Julian Huxley had been an ardent Bergsonian. His first book, *The Individual in the Animal Kingdom* (1912) was effusive in acknowledging the debt to Bergson. "It will easily be seen how much I owe to M. Bergson, who, whether one agrees or no with his views, has given a stimulus (most valuable gift of all) to Biology and Philosophy alike" (vii–viii). Bergson is praised for telling us that "that in any consideration of that system, it is the unity of it as a whole that is important: more than that, even if you want to consider a part of the system by itself, you cannot do so, for it loses almost all its significance when detached from the whole" (9). Echoing Spencer: "Bergson points out the inner unity for the good of which that action is performed. From the latter we can deduce another attribute of individuality, its heterogeneity; from that very unity of the whole we can postulate diversity of its parts" (9–10). Organicism reigns! "Bergson somewhere makes the illuminating remark that the whole of Evolution might have realized itself in a single individual" (20).

The inevitability of biological progress became the leitmotif of all of Huxley's subsequent writings, despite the fact that he tried to remain, at least in part, true to Darwinian mechanism. In *The Individual in the Animal Kingdom* (1912), he drew an interesting cultural-biological analogy, using the turn-of-the-century naval arms race between Britain and Germany as the example. First, the cultural: "Halfway through the century, when guns had doubled and trebled their projectile capacity, up sprang the 'Merrimac' and the 'Monitor,' secure in their iron breast-plates; and so the duel has gone on" (115). Concluding: "Each advance in attack has brought forth, as if by magic, a corresponding advance in defence." Then, the biological: "With life it has been the same: if one species happens to vary in the direction of greater independence, the inter-related equilibrium is upset, and cannot be restored until a number of competing species have either given way to the increased pressure and become extinct, or else have answered pressure with pressure" (115). Adding: "So it comes to pass that the continuous change which is passing through the organic world appears as a succession of phases of equilibrium, each one on a higher average plane of independence than the one before, and each inevitably calling up and giving place to one still higher" (116).

For a Darwinian, technically, there should be no absolute values here. The progress is relative. Having greater weapons of attack than your opponents might backfire. Imagine yourself in a conflict where fear and trembling—with an inclination to get out of here as soon as possible—is a far better strategy. Think of Hitler and Operation Barbarossa (the failed German attempt from 1941 to invade and conquer the Soviet Union). No such doubts or qualifications troubled Huxley, as one might expect from one who saw Spencer-style that "the organic world appears as a succession of phases of equilibrium, each one on a higher average plane of independence than the one before, and each inevitably calling up and giving place to one still higher." Absolute value-laden progress all the way, leading to humankind. "One somewhat curious fact emerges from a survey of biological progress culminating in for the evolutionary moment in the dominance of Homo sapiens. It could apparently have pursued no other course than that which it has historically followed" (Huxley 1942, 569). Little wonder that more than one person wondered about Huxley's total commitment to natural selection. To the amazement, shock, and dismay of secular biologists, the atheist Huxley was the president of the British Teilhard de Chardin Society, celebrating the works of the French Jesuit Bergson-influenced paleologist. In Chapter 8, we will take up Teilhard's thinking in its own right. For now, it is enough to learn that Teilhard argued that evolution is a progression up to the Omega Point, that he identified with Jesus Christ.

The Bergsonian influence was long-lasting. True, Huxley made much of natural selection. He also made much of other mechanisms, notably Sewall Wright's genetic drift. Almost cruelly, he applied this to the phenomenon that convinced Darwin of evolution, the differences between clearly related organisms on oceanic islands: "while geographical divergence always depends for its initiation on special isolation, it may subsequently be linked in varying degrees with ecological divergence of an adaptive nature, and also, in small populations, with non-adaptive divergence due to the genetic accident of 'drift.'" Huxley also made much of organization, not always to the credit of selection. For humans, and apparently for birds, "the upper limits of his powers and aptitude of mind were not determined by the struggle for existence" (534, quoting Harrison 1936). The demands of "organization"—putting everything together in a functioning whole—are the final arbiters. Likewise, "given the complex emotional make-up of song-birds, song is uttered in many circumstances where it has other functions or is even functionless, produced for 'its own sake.'"

Little wonder that, Spencer-like, Huxley felt free to use his biology as the jumping off point for a philosophical foray into the nature and foundations of morality. He argued that evolution justifies an obsession with technology, science, and major public works: "the individual is meaningless in isolation, and the possibilities of development and self-realization open to him are conditioned and limited by the nature of the social organization. The individual thus has duties and responsibilities as well as rights and privileges, or if you prefer it, finds certain outlets and satisfactions (such as devotion to a cause, or participation in a joint enterprise) only in relation to the type of society in which he lives" (Huxley 1934, 138–139). Reflecting this philosophy, at the same time as his *Synthesis* book, Huxley wrote a book on the Tennessee Valley Authority, which brought electricity all over the American South (Huxley 1943). After the Second World War, he was the first director general of UNESCO, hoping through that organization we would better the world, and he insisted that the S—science—be added to the name (Huxley 1948). His enthusiasm for progress so upset his conservative sponsors that his tenure was clipped from four years to two (Clark 1960), but that didn't change his beliefs in progress for all.

ORGANIZATION

For Schelling and his successors, organization is all important. Note that this is not just a matter of biology, for—as Spencer argued about dynamic equilibrium—the forces are as much physiochemical as they are biological. The early-twentieth-century Scottish morphologist D'Arcy W. Thompson thought this way. "Cell and tissue, shell and bone, leaf and flower, are so many portions of matter, and it is in obedience to the laws of physics that their particles have been moved, molded and conformed . . . Their problems of form are in the first instance mathematical problems, their problems of growth are essentially physical problems, and the morphologist is, *ipso facto,* a student of physical science" (1917, 10). D'Arcy Thompson is long gone but not forgotten. Those who work and write in this mode today are often those who come to biology, not from the dissecting lab but rather from the computer lab. They think their models can generate all that is needed—natural selection, take a holiday! In the words of theoretical biologist Stuart Kauffman: "The tapestry of life is richer than we have imagined. It is a tapestry with threads of accidental gold, mined quixotically by the random whimsy of quantum events acting on bits of nucleotides and crafted by selection sifting. But the tapestry has an overall design, an architecture, a woven cadence and rhythm that reflects underlying law—principles of self organization" (1995, 185; see also Kauffman 1993, 2008). Self-organization, a favorite term of Schelling. The pre-Socratic philosopher Parmenides argued that all change is illusory. Perhaps he is right, and nothing changes.

Canadian-born Brian Goodwin was a biologist, and although he was a professional maverick, no one doubted his abilities—or genuine concern to discover the truth. Striking right home at the Darwinians, Goodwin (2001) seized on supposedly paradigmatic examples of natural selection in action, arguing that in fact they are no more (no less either) than the unfurling patterns of nature itself—self-generating organization. Take phyllotaxis, the patterns shown by many flowers and fruits in the plant world, for instance the spirals of the sunflower or the twisting lines shown by pinecones. Darwinians had long argued that these are of direct adaptive significance, usually associated with maximizing the amount of sunlight that falls on some specific leaf or other. Goodwin, instead, seized upon the mathematics of the case. The ways of growth force the components into certain familiar grids or lattices, and these in turn are amenable to simple mathematical analysis. The plants in question produce their parts from the center and then push out as they grow. In a sunflower, for instance, one gets one leaf and then another and then another—this produces the genetic spiral at the heart of the flower. As the leaves line up, one by one, new lines or patterns emerge—the most noticeable spirals are known as parastichies. The leaves running along any particular spiral, numbering them in the order they were produced, exhibit fixed patterns. Remarkably the numbers from the

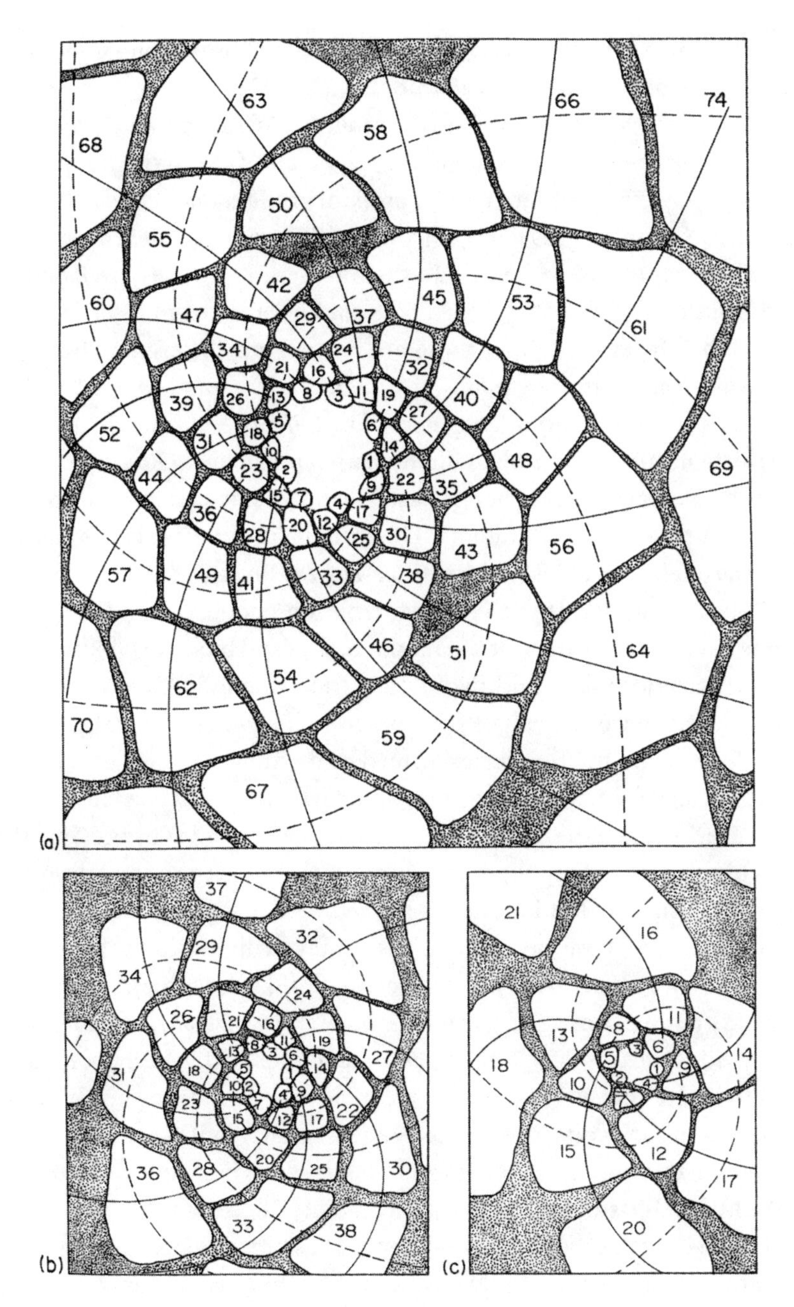

FIGURE 2.2 Phyllotaxis as shown by the monkey puzzle tree. *Credit:* Republished with permissions of Princeton University Press from Brian Goodwin, *How the Leopard Changed its Spots*, © 2001; permission conveyed through Copyright Clearance Center, Inc.

crisscrossing parastichies have a formula. Goodwin uses as an example the leaves of a monkey puzzle tree (Figure 2.2).

The numbers on the leaves are the order in which they were produced, with 1 being the most recent, and then 2, on out to 74 in the top picture (a). Now observe the differences (in order of production) between the spirals: the touching leaves going the one way (clockwise) and those going the other way (anti-clockwise). In the top picture (a), they are 8 and 13. In the bottom left (b), they are 5 and 8, and in the bottom right (c), they are 3 and 5. It turns out that these are not random differences between the two spirals in any one plant. They follow the sequence: 0, 1, 1, 2, 3, 5, 8, 13 . . . This is the formula—general form, $n_i = n_{i-1} + n_{i-2}$—worked out by the thirteenth-century Italian mathematician Leonardo Fibonacci, who tried to calculate the number of descendants in any generation from an initial couple of breeding rabbits. The formula is of course better known today as one of the clues in the thriller *The Da Vinci Code*.

Goodwin argued that that is explanation enough. The developing plants simply follow the rules of mathematics, and biological forces (like natural selection) have nothing to do with it. Goodwin's ideas were not just Platonic but practically Pythagorean in his numerological enthusiasms. The vulgar fraction series formed by dividing successive members of the Fibonacci series homes in on 0.618, which in turn is what the Ancient Greeks called the Golden Mean, the figure arrived at by dividing the sides of a rectangle such that removing a square from the rectangle leaves one with a smaller but identically shaped rectangle. As it happens, you can get the Golden Mean out of circles too if you divide up the perimeter in a certain way. This gives you a major angle of 137.5 degrees, which (and if you are not yet convinced you will be now!) is just the angle on the spiral that divides successive leaves or parts. "So plants with spiral phyllotaxis tend to locate successive leaves at an angle that divides the circle of the meristem in the proportion of the Golden Mean. Plants seem to know a lot about harmonious properties and architectural principles" (Goodwin 2001, 127). (Meristems are the growing points of plants.)

Following a path for which our history has prepared us, for Goodwin, all of this reveals a philosophy that sees everything as interconnected, in an essentially harmonious fashion, with shared values. As a good Platonist, Goodwin had nothing but contempt for a philosophy that attempts to take meaning and value out of existence. Darwinism is "an extreme reductionism that makes it impossible for us to understand concepts such as health. Health refers to wholes, the dynamics of whole organisms. We currently experience crises of health, of the environment, of the community. I think they are all related. They are not caused by biology by any means, but biology contributes to these crises by failing to give us adequate conceptual understanding of life and wholes, of ecosystems, of the biosphere, and it's all because of genetic reductionism" (King 1996, 6–8). We have got to escape the Darwinian metaphors of "competition and conflict and survival," replacing them with metaphors stressing organisms as "co-operative as they are competitive." We must turn from "nature red in tooth and claw, with fierce competition and the survivors coming away with the spoils." We need a new perspective where the "whole metaphor of evolution, instead of being one of competition, conflict and survival, becomes one of creativity and transformation."

SOCIOBIOLOGY

Now let us look at a specific instance of how, more recently, Darwinian evolutionists took their newly confident paradigm, which (if Kuhn is right) had been turned now to normal science, as they started to explore and extend their base. This takes us at once to a remarkable clash between leading evolutionists, a clash that turned out to have depths unsuspected at the time. In this debate, we shall show the persistence of the organic perspective, even after the supposed triumph of Neo-Darwinism.

As the twentieth century approached its middle point, evolutionary biologists started to turn their attention to social behavior. Pathbreaking were the continental "ethologists," such as the Austrian Konrad Lorenz and the Dutch Niko Tinbergen. As the centenary of the *Origin* approached (1959), more and more anglophones became interested, although perhaps expectedly given the continental influence, in a somewhat non-Charles-Darwinian fashion, the main hypothesized causal process was group selection. It was argued that if we are to understand animals—insects, mammals, birds—acting socially, then we must suppose that selection works on them as groups rather than individually. Then, at the beginning of the 1960s, something truly revolutionary did happen—not revolutionary in throwing over the neo-Darwinian paradigm, but revolutionary in the sense of causing a major change in the study and understanding of the evolution of social behavior. A major change emphasizing the mechanistic, reductionistic nature of the science. Thanks particularly to the then-graduate student William Hamilton (1964a, 1964b), hardline individual selection explanations—that is, reductionistic explanations—were offered for the sociality of the Hymenoptera, an order of insects that includes wasps, bees, and ants.

Darwin-fashion, Hamilton pointed out that all the female inhabitants of the nest are related, but thanks to the peculiarities of hymenopteran ways of reproduction, sisters are more closely related to one another (so long as their mothers are monogamous) than are mothers to daughters. Males in the Hymenoptera have only one set of chromosomes, whereas females have the usual two sets. When the queen's eggs are fertilized, they become females and generally have the role of workers, but if the eggs are not fertilized, they develop nevertheless and become males, the drones. This means that female full sisters are 75 percent related to each other, whereas mothers and daughters are 50 percent related. Hence, being a sterile worker and helping your fertile sisters is going to mean more of your genes are passed on than if you were fertile. Of course, fertile sisters, queens, do well out of the system, but most females cannot become queens. Males don't have such an asymmetry, and so, as expected, sterility is not a career option.

The beauty is that this is individual selection right down the line. Workers are not sacrificing themselves for the good of the group. They are looking after their own interests by reproducing by proxy as it were. Although the Hymenoptera are a special case, this process of reproducing by proxy—labeled (a term already introduced) "kin selection" by one of Hamilton's PhD examiners, John Maynard Smith (1964)—is widely applicable. If resources are limited, it might be in the interests of weaker members of the family to help out their existing relatives than to go it alone. As applicable are other posited individual selection-based sociality-promoting mechanisms. Darwin himself spotted one: "as the reasoning powers and foresight of the members [of a tribe] became improved, each man would soon learn from experience that if he aided his fellow-men, he would commonly receive aid in return" (Darwin 1871, 1, 163). Reciprocal altruism. You scratch my back, and I will scratch yours. And with tools like these, it was not long before evolutionists started to go out into nature and look now at social behavior from a Darwinian—individual selection—perspective, with much success on animals ranging in size from elephants to dung flies.

E. O. WILSON AND HIS CRITICS

Enter Edward O. Wilson, Harvard professor, expert on the ants and methods of chemical communication and more broadly co-author of some of the most noted quantitative work on biogeography (MacArthur & Wilson 1967). Now he turned his hand to the evolution of social behavior, increasingly known as "sociobiology," authoring a massive, beautifully illustrated tome on the subject. *Sociobiology: The New Synthesis* (1975)—deliberately the title echoes Julian Huxley's *Evolution: The Modern Synthesis*—is breathtaking in its scope. It gives theoretical discussion followed by a sweeping survey from colonial invertebrates, through social insects, on to mammals and then primates, culminating in great apes, and finally finishing with our species, *Homo sapiens*. Humans are part of the story. We dance to the genes like other animals.

The nuclear family is all-important. "The populace of an American industrial city, no less than a band of hunter-gatherers in the Australian desert is organized around this unit. In both cases, the family moves between regional communities, maintaining complex ties with primary kin by means of visits (or telephone calls and letters) and the exchange of gifts. During the day the women and children remain in the residential area while the men forage for game or its symbolic equivalent in the form of barter and money" (Wilson 1975, 553). And so on: "sexual bonds are carefully contracted in observance with tribal customs and are intended to be permanent. Polygamy, either covert or explicitly sanctioned by custom, is practiced predominantly by the males" (553–554). What about outliers? What of homosexuals for instance? Kin selection to the rescue. "The homosexual members of primitive societies may have functioned as helpers, either while hunting in company with other men or in more domestic occupations at the dwelling sites. Freed from the special obligations of parental duties, they could have operated with special efficiency in assisting close relatives" (555).

Not all scientists agreed with Wilson. Richard Lewontin, geneticist and Marxist, Wilson's fellow department member at Harvard, was scathing. In an address, "Sociobiology—a caricature of Darwinism," given the year after *Sociobiology* was published, at the end of a long list of complaints, he concluded:

> Finally sociobiological theory rests on an erroneous confusion between materialism and reductionism. It is sure that we are material beings and that our social institutions are the products of our material beings, just as thought is the product of a material process. But the content and meaning of human social organization cannot be understood by a total knowledge of biology any more than by a total knowledge of quantum theory. War is not the sum total of individual aggressive feelings and a society cannot be described if we know the DNA sequence of every individual in it. The naive reductionist program of sociobiology has long been understood to be a fundamental philosophical error. Meaning cannot be found in the movement of molecules. (Lewontin 1976, 2, 31)

Materialism but not reductionism.

Was Lewontin's critique purely philosophical, the commitment of a Marxist to early-nineteenth-century Germanic holistic philosophies? It was certainly all this. Against "reductionistic explanation," we learn: "Dialectical explanations, on the contrary, do not abstract properties of parts in isolation from their associations in wholes but see the properties of parts as arising out of their associations" (Lewontin et al. 1984, 11). However, Lewontin would have added indignantly that his critique was as much scientific as philosophical. Along with fellow evolutionist, fellow member of the same Harvard department as Wilson, fellow Marxist Stephen Jay Gould, Lewontin authored a celebrated article, "The Spandrels of San Marco and the Panglossian Paradigm: A Critique of the Adaptationist Programme." Lewontin made clear his organismic approach to the problems of biology. He, with Gould, argued against the "faith in the power of natural selection as an optimizing agent," something that "proceeds by breaking an organism into unitary 'traits' and proposing an adaptive story for each considered separately" (Gould & Lewontin 1979). They wanted "to reassert a competing notion (long popular in continental Europe) that organisms must be analyzed as integrated wholes, with bauplāne [underlying morphological archetypes] so constrained by phyletic heritage, pathways of development, and general architecture that the constraints themselves become more interesting and more important in delimiting pathways of change than the selective force that may mediate change when it occurs" (581).

Here, we turn back to Goethe. "In continental Europe, evolutionists have never been much attracted to the Anglo-American penchant for atomizing organisms into parts and trying to explain each as a direct adaptation" (Gould & Lewontin 1979, 159). So, what is the alternative? It is one that "acknowledges conventional selection for superficial modifications of the *Bauplan.* It also denies that the Adaptationist Programme (atomization plus optimizing selection on parts) can do much to explain *Baupläne* and the transitions between them. But it does not therefore resort to a fundamentally unknown process. It holds instead that the basic body plans of organisms are so integrated and so replete with constraints upon adaptation . . . that conventional styles of selective arguments can explain little of interest about them" (594). Natural selection's existence is somewhat begrudgingly acknowledged, but more as a clean-up process after the real action has occurred: "constraints restrict possible paths and modes of change so strongly that the constraints themselves become much the most interesting aspect of evolution" (594).

As is usually the case with these sorts of clashes, the divide between Wilson and his critics was not just epistemological. Ethical issues were close to, if not right on, the surface. In the opinion of Gould, Lewontin, and other critics, theories such as Wilson's "consistently tend to provide a genetic justification of the *status quo* and of existing privileges for certain groups according to class, race or sex." Continuing: "He purports to take a more solidly scientific approach using a wealth of new information. We think that this information has little relevance to human behavior, and the supposedly objective, scientific approach in reality conceals political assumptions. Thus, we are presented with yet another defense of the status quo as an inevitable consequence of 'human nature'" (Allen et al. 1975, 43).

DIGGING BENEATH THE SURFACE

And now comes the not-then-recognized paradox at the heart of all of this. Edward O. Wilson was as much a Spencerian organicist as any of them! He may come across as a Darwinian mechanist, but truly his allegiances lay elsewhere (Gibson 2013). As noted already, in the early twentieth century at Harvard, many of the biologists were Spencer enthusiasts. This applied especially to those working in the field of social behavior. They believed strongly in group processes, including selection, and they saw life as moving upwards, from equilibrium point to equilibrium point, including, especially including—as for Spencer—humankind. A good example is William Morton Wheeler (Evans & Evans 1970), the intellectual grandfather of Edward O. Wilson. (He was the supervisor of Wilson's supervisor.) He was enthusiastic about analogies between human and ant societies, going on to speak of ant societies as being akin to individual organisms. To be candid, Wheeler was not overly enthusiastic about natural selection. Like Spencer, he emphasized cooperation as an important aspect of organic life as opposed to struggle. Wheeler was much interested in the implications of all of this for our human society, as were Harvard colleagues such as Walter B. Cannon (1931) and Lawrence J. Henderson (1917), and (reflecting his own area of expertise) did not feel unduly optimistic. "Given the atrophy or subatrophy of our organs and tissues brought about by the ever-increasing specialisation in our activities, unfortunately we can hardly fail to suspect that the eventual state of human society may be somewhat like that of the social insects—a society of very low intelligence of the individuals combined with an intense and pugnacious solidarity of the whole" (Wheeler 1939, 162). Although Darwin treated the ant colony as an individual, Wheeler went way beyond Darwin in thinking of a human society as akin to an ant colony. Wheeler was prepared to treat a group of unrelated humans as a unit of selection in a way that Darwin would never have done. For Darwin, it was the relatedness that counted; for Wheeler, it was the group.

Wilson, born in 1929, came to scientific maturity in the 1950s, just at the time when mechanistic reductionistic science was at its most forceful. Later in that decade, one of Wilson's colleagues at Harvard was James Watson of double-helix fame and a strong advocate of a molecular approach to nigh everything. Although, notoriously, Wilson (1994) described Watson as the most unpleasant person he ever knew, he felt the reductionist influence and, in the 1960s, became an enthusiastic convert to Hamilton's thinking on the Hymenoptera and of kin selection generally. This is reflected in his discussion of social behaviors, altruism particularly, in *Sociobiology* (1975), although with hindsight, Wilson was already having second thoughts. The discussion in fact is remarkably sympathetic to group selection. It is not ruled out a priori, and indeed some forms seem accepted. Having introduced the idea of kin selection, Wilson wrote: "Selection can also operate at the level of species or entire clusters of related species. The process, well known to paleontologists and biogeographers, is responsible for the familiar patterns of dynastic succession of major groups, such as ammonites, sharks, graptolites, and dinosaurs through geologic time" (106). This said, when it comes to altruistic behavior between individuals, Wilson steps sharply into line. "However, selection at these highest levels is not likely to be important in the evolution of altruism, for the following simple reason. In order to counteract individual selection, it is necessary to have population extinction rates of comparable magnitude. New species are not created at a sufficiently fast pace to be tested in this manner" (106).

In subsequent work on the social insects, however, Wilson, in the opinion of many of his fellow professional evolutionists, went over to the dark side. He showed strong opposition to the whole individual selection *über alles*—selfish-gene—line of thought and argued that one must regard nests as individuals, "superorganisms," units of selection in their own right (Hölldobler & Wilson 2008; Wilson & Wilson 2007; see also the already-mentioned Sober & Wilson 1998). Cooperation leads to "emergent properties" favoring the group. "Consider genetic variation of traits such as nest construction, nest defence, provisioning the colony for food, or raiding other colonies. All of these activities provide public goods at private expense. All entail emergent properties based on cooperation among the colony members. Slackers are more fit than solid citizens within a single colony, but colonies with more solid citizens have the advantage of the group level" (Wilson & Wilson 2007, 341).

And really, if you look again at *Sociobiology,* leaving on one side the friendly attitude toward group selection—he shocked purists by referring to kin selection as a form of group selection—there are many hints that Herbert Spencer is Wilson's true guide rather than Charles Darwin. Wilson tells us that of all animals: "Four groups occupy pinnacles high above the others: the colonial invertebrates, the social insects, the nonhuman mammals, and man" (Wilson 1975, 379). He continues: "Human beings remain essentially vertebrate in their social structure. But they have carried it to a level of complexity so high as to constitute a distinct, fourth pinnacle of social evolution" (380). He concludes by speaking of humans as having "unique qualities of their own." He now launches at length into showing us how humans have crossed over and mounted the "fourth pinnacle" (382)—the "culminating mystery of all biology" (382). All this, as Wilson makes clear in subsequent writings, is very much part of the general picture. "The overall average across the history of life has moved from the simple and few to the more complex and numerous. During the past billion years, animals as a whole evolved upward in body size, feeding and defensive techniques, brain and behavioral complexity, social organization, and precision of environmental control—in each case farther from the nonliving state than their simpler antecedents did" (Wilson 1992, 187).

If this isn't an organicist picture of life's history, we don't know what is. We doubt that Wilson has even heard of Friedrich Schelling, let alone read him, but the tradition lives on. Moreover, for all that Lewontin and company would claim the higher moral ground, Wilson would dispute this bitterly. His philosophical-scientific background is the basis for a lifelong passionate environmentalism. This is at the heart of his "biophilia" hypothesis. "To explore and affiliate with life is a deep and complicated process in mental development. To an extent still undervalued in philosophy and religion, our existence depends on this propensity, our spirit is woven from it, hope rises on its currents" (1984). For the mechanist, although nature is surely something of great value—who could view the Canadian Rockies in the early morning without awe, an overwhelming sense of beauty and worth?—it is value you ascribe rather than find. In the organicist tradition, for Wilson, nature itself is impregnated with value. All of nature, for life is an interconnected whole. Individuals—organisms or species—are part of a larger network. No one or group can exist apart in isolation. Morally, therefore, if we care about humans we should care about all life.

TWO APPROACHES

As we move toward the present, the topic of Chapter 3, let us contemplate this arc from organic worldview to mechanistic and back and forth. When considering this great debate, one might think that since Darwin and the *Origin,* in evolutionary studies, we have seen the triumph of the mechanistic / reductionistic approach to scientific questions. At one level, we do not dispute this. If Hamilton's work on the Hymenoptera (ants, bees, and wasps) is not mechanistic / reductionistic, we don't know what would be. However, there is another level.

In evolutionary thinking, even before the time of Darwin, there has been a strong current of organicism. One sees this most obviously in those with more philosophical interests—Spencer, Bergson, and Whitehead—but although it would be a bad mistake to downplay the influence of these philosophers on the science of the day, one also sees a current of organicism in the highest levels of the science. Sewall Wright was without question one of the most important population geneticists of his day—his influence only equaled by that of Fisher. Not only was Wright himself pushing organicism, but his students followed suit. Dobzhansky ever had organicist yearnings, especially about progress, and as Huxley was the British president of the Teilhard de Chardin Society in Britain, Dobzhansky held the parallel role in America.

As a kind of codicil to this chapter, one of us (Ruse) was a leading figure, along with his fellow, rather-more-senior philosopher David Hull, in promoting, in the 1960s, what one might call the "Renaissance" of the philosophy of biology. New ideas were offered, and new theories supposed. Ruse wrote one introductory overview—*The Philosophy of Biology* (1973)—and Hull wrote another—*The Philosophy of Biological Science* (1974). Sparking this movement, in major part, was the fact that, as we have seen, by the 1960s, biology, particularly evolutionary biology, was moving forward confidently, with the basic outlines of Neo-Darwinism (the synthetic theory) now firmly sketched out and incorporating new ideas and models, such as William Hamilton's work on kin selection. Also obviously important were the major developments in molecular biology, by then—as we shall learn, rather amusingly, with the work of one Richard Lewontin—starting to have major consequences for more conventional biology. Another significant factor in the growth of the philosophy of biology, however, was the strong feeling that, to this point, although the philosophy of physics had made major strides in the past half century, wrestling with such major advances as relativity theory and quantum theory, the philosophy of biology had been stagnant, if not worse. To call what then existed "second rate" was to exaggerate. Ruse and Hull, ardent pathfinders, were scathing in their assessments. Yet today, an increasing number of historians and philosophers of science think theirs was a totally unjustified belittling of the work in the philosophy of biology that had been produced in the fifty or more years earlier. In a review of a book to which Michael Ruse contributed, the reviewer states bluntly:

> The last issue that deserves to be mentioned is Michael Ruse's self-proclaimed paternity over the philosophy of biology and his renewed attack on early-twentieth-century philosophers of biology, particularly Joseph H. Woodger. Without wanting to downplay Ruse's tremendous contributions to the philosophy of the life sciences, I cannot help pointing out that his depiction of early-twentieth-century philosophy of biology as "little literature [. . .] that is almost uniformly bad" and that "left much to be desired" (p. 277) has been recently put into serious question by the in-depth historiographical work of Nicholson and Gawne (2014; 2015; see also Peterson 2016). Moreover, some central research topics in present-day philosophy of biology pay a considerable debt to and stand in continuity with the early-twentieth-century philosophy of biology that Ruse belittles (see Nicholson 2014; Baedke 2019). (Prieto 2021, 615)

In response, we authors agree that Ruse was very much mistaken! However, we feel that his mistake is of the kind that is more constructive than destructive, since it points to an interesting and relevant reason why Ruse was mistaken. Note whom the reviewer picks out in refutation—Joseph H. Woodger. To him could be added the Whiteheadians as well as others, for instance the Austrian biologist Ludwig von Bertalanffy. What is distinctive about all these people? To a person, they were organicists! In his *Biological Principles* (1929), Woodger writes of "the 'organic view' of nature," acknowledging that it "is championed by Professors Whitehead, Lloyd Morgan and others." He writes of its "importance from the biological point of view in its emphasis on the concept of organization, and on relations—especially internal and multiple relations," adding that "in this book I have tried to analyse the notion of organization in relation to the actual problems of biology (a task which, so far as I know, has not been attempted by biologists), with the aids provided by modern thought, but here again I do not profess to have done more than open up the subject" (484–485). Ruse and Hull were simply not sensitive to these sorts of issues—organicist issues. They approached biology from the perspective of mechanism, and had they been challenged, they would have said that twentieth-century biology—Mendelian genetics, population genetics synthesizing Darwinian selection with the new concept of the gene, molecular biology especially the double helix, kin selection, and more—justifies their approach. Our point today is not that Ruse and Hull were wrong, rather to note how they worked from their perspective and simply could not see the virtues of the other perspective. Exactly the kind of situation we have been sketching in this chapter.

It is interesting to note that while Ruse stayed firmly within the mechanistic perspective with the machine as the explanatory root metaphor, in his later writings, Hull (1976, 1978) moved decisively toward organicism, especially in his championing of the "species as individuals" thesis, seeing—a position endorsed by Woodger—biological species as integrated wholes rather than sets of individual organisms. Hull's move between metaphors reinforces what we have seen before in our journey through the century. For all the purported discoveries within the mechanist worldview, as noted at the beginning of this chapter, organicism proves a hardy flower. Richard Lewontin, Dobzhansky's student, was overtly critical of blind mechanism and reductionism. Rather deliciously, the chief target of his philosophical and scientific ire, Edward O. Wilson, who died in 2021, was firmly embedded in the organicist movement. The point we make is that as we move to the present, although we shall see new (or revived) motives and innovative strategies, there is a long, distinguished—high class—tradition of organicism. Today's organicists are carrying on the tradition, not starting a new one. Truly, the "New" Biology is not so very new after all!

Chapter Three

THE NEW BIOLOGY

> It is futile to argue whether reductionism is wrong or right. But this one can say, that it is heuristically a very poor approach. Contrary to the claims of its devotees, it rarely leads to new insights at higher levels of integration and is just about the worst conceivable approach to an understanding of complex systems. It is a vacuous method of explanation.
>
> —MAYR 1969, 128

THE NEO-ARISTOTELIANS

Expectedly, as we prepare the way for today's scientific debate, there are various cultural debates that could be used to ease entry (Bateson et al., 2017). Debates that show scientists who today want to push an organic perspective are not oddballs, outliers, but part of a strongly supported past and present way of approaching the world. Two of the cultural debates, the Gaia hypothesis and ecofeminism, will be discussed later (in Chapter 7). Here, let us give the floor to the philosophers and the renaissance of an Aristotelian view of the world that has occurred over the past half century. Most notable is the area dealing with moral thinking and behavior, ethics, and the revival of so-called Aristotelian virtue theory (Hursthouse 1999). These theories emphasize the ideal character and virtue in a person's response to situations. For example, what would a virtuous person do if a child was crying? They would help the crying child. Likewise, they would refrain from pushing someone under a bus "just for kicks." Virtue ethics is teleological, focusing on ends. The aim is to achieve a kind of enlightened state of well-being: *eudaimonia,* sometimes translated as "flourishing." The final cause, the purpose of human life, is to use reason to achieve this state. One does not act randomly or instinctively. One has certain developed characteristics known as "dispositions," a disposition to be compassionate, for instance. Aristotle refers to these as "virtues"—not so much emotions, but appropriate psychological dispositions to specific situations. One is guided by (a philosophical reading of) the Golden Mean—thus, for the emotion of fear, the vice of deficiency is cowardness, the virtuous disposition exhibits courage, and the vice of excess is recklessness. The point is that, for Aristotle, morality is not a matter of acts as such but rather of character that will lead one to acts—rescuing a wounded soldier in no-man's land using what cover there is—that are virtuous as opposed to cowering in the trenches or dashing wildly toward enemy lines exposing oneself to no good effect. Everything is to be understood in terms of purposes or ends.

It is natural to extend this kind of Aristotelian thinking about ends from ethics, theory of morality, to epistemology, theory of knowledge. One who does this with enthusiasm is the eminent American analytic philosopher, Thomas Nagel. He makes no bones about where he stands, having authored *Mind and Cosmos: Why the Materialist Neo-Darwinian Conception of Nature Is Almost Certainly False* (2012). Nagel is at one with Darwin in finding the design-like nature of the organic world overwhelming. However, what sticks in his craw are the random, undirected elements in the Darwinian picture. If this be the machine-metaphor perspective, so much for the machine-metaphor perspective. Nagel opines that possibly "there are natural teleological laws governing the development of organization over time, in addition to laws of the familiar kind governing the behavior of the elements." He allows that "This is a throwback to the Aristotelian conception of nature, banished from the scene at the birth of modern science. But I have been persuaded that the idea of teleological laws is coherent, and quite different from the intentions of a purposive being who produces the means to his ends by choice. In spite of the exclusion of teleology from contemporary science, it certainly shouldn't be ruled out a priori" (22). As you may have recognized, what is at work here is the organicist model—direction to an end, acorn to oak tree, monad to man—rather than the mechanist model. (See also Dupré 2010, 2012a, b, 2017; Fodor 2007; Sterelny 2012.)

Time to turn to the science. As we do so, let us emphasize—whether one agrees with the science, or one decides that Darwinism has run its route—is one thing. This in no sense prejudices the other thing, the social motives behind the science. They are important for the context of the science, and they are powerful.

THE EXTENDED EVOLUTIONARY SYNTHESIS

One point of clarification as we move forward. Up to this point, we have been using the term "paradigm" to refer to something like a scientific theory, as in the "Darwinian paradigm." The term "paradigm" is notoriously slippery, and—where there is no ambiguity—from now on, rather than speaking of the mechanist perspective or worldview, we shall feel free to use it in the broader sense of "world picture" or "root metaphor," as in the "mechanist paradigm." We make this move to stress that the mechanism / organicist debate is very much one of commitments, as in politics and religion, and not just the "facts," whatever these might be. Obviously, this extension of meaning is not a major move, since Darwinian theory is the epitome of a mechanistic theory or working within the machine root metaphor, and the countersuggestions are the epitome of organist theories or working within the organic root metaphor.

In the past decade, a number of "New Biologists"—our organismic evolutionists or, as they label themselves, "Extended Evolutionary Synthesis" (EES) supporters—have been holding symposia and publishing position papers, sometimes alone and sometimes with Neo-Darwinian mechanists—"Standard Evolutionary Theory" (SET) supporters—aiming at clarifying differences and promoting their position. "We see the version of evolutionary theory that is being advocated today, commonly called the 'Extended Evolutionary Synthesis' (EES), as an updated version of the early twentieth-century organicists' evolutionary view on a higher plane, growing from, yet also challenging, the Modern Synthesis"

(Jablonka & Lamb 2020, 1). Or as one might put it: "The EES is not a simple, unfounded call for a new theory but has become an ongoing project for integrating the theoretically relevant concepts that have arisen from multiple fields of evolutionary biology" (Müller 2017, 8). As a consequence, "the EES establishes a new structure of the theoretical evolutionary framework that goes beyond the reductionist and gene-centered perspective of the past. It represents a different way of thinking about evolution, historically rooted in the organicist tradition. Its predictions permit the derivation of new hypotheses and thus inspire novel and progressive research in evolutionary biology and adjacent fields" (8). The New Biologists have three main points of discontent with old ways of thinking (SET). The first focuses on evolutionary development biology. The second on developmental plasticity. The third on niche construction theory. We take them in turn.

When SET was put together in the 1930s, the tacit agreement was to leave untouched the move from genotype to phenotype. Obviously, it was important, but the main synthesizers were not embryologists and, with some few exceptions—Gavin de Beer (1940) was one—embryologists were not that interested in evolution. Study of development was something that could be done in isolation in the lab, and in any case, homologies rather than adaptations were the chief interest. For Darwin and his followers, homology was important but always secondary to adaptation. Hence, development was something of a "black box." Genes in, organisms out.

Molecular biology was obviously the tool to open up the "black box"—once you knew how the nucleic acids could produce proteins, the crucial components of living beings, you were already on your way to looking at development. So grew up the subfield of evolutionary development or "evo-devo," although note that you are still looking at things rather in isolation—less what function does it have and more how does it work. You might think that this is all very mechanistic, and we shall see that the mechanists argue that it is. The evo-devo enthusiasts likewise admitted this. "Evo-devo provides a causal-mechanistic understanding of evolution by using comparative and experimental biology to identify the developmental principles that underlie phenotypic differences within and between populations, species and higher taxa" (Laland et al. 2015, 3). But for those of an organismic bent, this was just the background (Laland et al.

2014). The key is internal development rather than being guided exclusively by external forces. "While much evo-devo research is compatible with standard assumptions in evolutionary biology, some findings have generated debate. Of particular interest is the observation that phenotypic variation can be biased by the processes of development, with some forms more probable than others" (Laland et al. 2015, 3). Numbers of digits and vertebrae are instanced as supportive of this claim.

Likewise, the similarities one finds between different species. SET supporters would argue for convergent evolution—exemplified by the saber-toothed tigers that had placental and marsupial representatives, both utilizing similar opportunities. EES supporters invite you to take the example of similar cichlid fishes in different lakes in Tanganyika. "Such repeated parallel evolution is generally attributed to convergent selection. However, inherent features of development may have channeled morphology along specific pathways, thereby facilitating the evolution of parallel forms in the two lakes" (Laland et al. 2015, 3). "Channeling" is a popular word here, but the important point is that the channeling leads to adaptive features, without need of natural selection. It all comes through the flowering of the organism.

The second point of discontent was developmental plasticity, "the capacity of an organism to change its phenotype in response to the environment" (Laland et al. 2015, 3). Today, it is openly—and comfortably—admitted that this is not entirely a new idea. Back at the time of the forming of SET, English evolutionist C. H. Waddington was arguing for a kind of genetic Lamarckism. Organisms would respond to the environment—the blacksmith getting stronger arms—and then somehow this change would get encoded in the genes. "Genetic assimilation," as Waddington (1953, 1957) called it, would lead to directed mutations in some sense. He subjected fruit flies to extreme, shocking conditions, causing anomalies in growth, and then he selected from the anomalies. Within a very few generations, he got the anomalies without need of environmental disturbance, the shocks. What made Waddington's work particularly controversial and unwelcome was that these kinds of claims were being made when mainstream Russian geneticists were under great pressure, thanks to the endorsement by Stalin of Lysenko and his Lamarckian-type theories. This, in fact, had nothing to do with Waddington's motivation. He was an ardent follower of A. N. Whitehead, and his work was entirely within the organicist paradigm. He was trying to show how variations are directed as that perspective supposes. The New Biologists echo Waddington. "It has long been argued that phenotypic accommodation could promote genetic accommodation if environmentally induced phenotypes are subsequently stabilized and fine-tuned across generations by selection of standing genetic variation, previously cryptic genetic variation or newly arising mutations" (Laland et al. 2015, 4). They add that this process means that evolution can be rapid as organisms take advantage of new opportunities.

The third main point of discontent was niche construction theory, "the process whereby the metabolism, activities and choices of organisms modify or stabilize environmental states, and thereby affect selection acting on themselves and other species. For example, many species of animals manufacture nests, burrows, webs and pupal cases; algae and plants change atmospheric redox states and modify nutrient cycles; fungi and bacteria decompose organic matter and may fix nutrients and excrete compounds that alter environments" (Laland et al. 2015, 4). Much of this sounds very much like the phenomenon the ultra-Darwinian Richard Dawkins (1982) labeled the "extended phenotype" (more on this notion in Chapter 7). Why, then, did it challenge or exceed SET? It is because niche construction can make the organism an active player in the process, directing the nature of selection, rather than sitting back passively and waiting for selection to do all the work, imposing itself upon the organism. Figure 3.1 makes the point. It could be just the environment (case 2) setting up the selective pressures on the organism. Here, the organism is a passive player, and it evolves under imposed natural selection. But if the organism is actively building or creating its own niche (case 1), then it can itself or through its activities change the selection pressures. Selection still plays a role, but the organism now is an active player.

Summing up:

> Niche construction initiates and modifies the patterns of natural selection directly affecting the constructor (and other species that share its environment) in an orderly, directed and sustained manner, in part because feedback [Figure 3.1] leads to a self-reinforcing process. As a consequence, niche

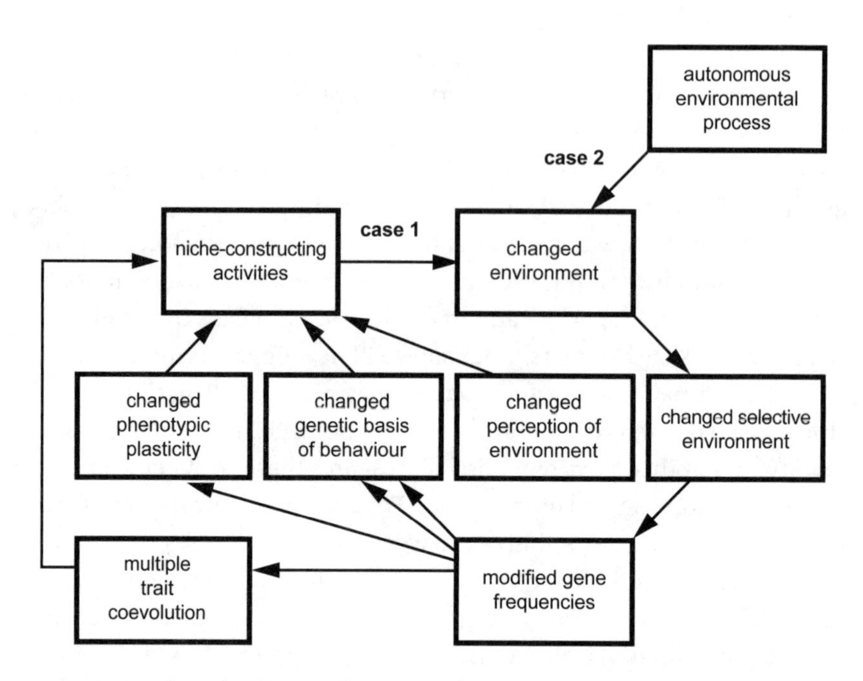

FIGURE 3.1 The cycle of cause-effect relationships associated with niche construction. The self-reinforcing nature of this cycle generates much less variation in the source of selection than where there is no feedback from organisms' activities to the environment. Here, "autonomous" refers to environmental processes that are not affected, or only weakly regulated, by the activities of organisms. Consequences of niche-constructed aspects of the environment (case 1) may be qualitatively different from environmental changes resulting from autonomous environmental processes (case 2), leading to experimentally detectable differences in selective environments, gene-frequency changes, and patterns of trait coevolution in the two cases. *Credit:* Reformatted with permission of The Royal Society (UK) from Kevin Laland, John Odling-Smee, and John Endler, "Niche construction, sources of selection and coevolution," *Interface Focus* 7:5, 2017, Figure 1; permission conveyed through Copyright Clearance Center, Inc.

> construction directs adaptive evolution. Niche construction should be recognized as an evolutionary process because it imposes a statistical bias on the direction and mode of selection that ensue, and hence on the speed and direction of evolution. By systematically creating and reinforcing specific environmental states, niche construction directs evolution along particular trajectories. (Laland et al. 2017, 2)

Attention is drawn to the long-term (forty years) studies by Peter and Rosemary Grant (2014) of the finches in the Galapagos. Candidly, the

Grants admit: "in the long-term evolution is unpredictable because environments, which determine the directions and magnitudes of selection coefficients, fluctuate unpredictably" (Grant & Grant 2002, 707). But with niche construction, things become much more controlled and predictable. "With niche construction there is feedback between the activities of organisms and the environment, such that the entire process can be self-reinforcing [Figure 3.1, case 1]. Like artificial selection, the direction of evolution is less subject to fluctuations than if the feedback were absent" (Laland et al. 2017, 3). And this means that "selection resulting from niche construction is likely to be qualitatively different from selection arising from autonomous (unaffected by organism's activities) environmental processes" (Figure 3.1, case 2). In niche construction, it is the organism that controls, not the environment.

DEFINING THEMES

The New Biologists, EES advocates, fully aware that the old biologists, SET believers, are going to regard their arguments with amusement if not derision—storms in teacups come to mind—defend their position from a more methodological / metaphysical perspective. "We see two key unifying themes to [our] interpretations—constructive development and reciprocal causation" (Laland et al. 2015, 6). Both themes stress the organismic nature of their approach, but for those who advocate the EES, "Constructive development refers to the ability of an organism to shape its own developmental trajectory by constantly responding to, and altering, internal and external states" (Laland et al. 2015, 6). Basically, what this means is that the organism is an active player in the evolutionary process, not sitting back passively letting natural selection do everything. Selection—and genetics—are coplayers in the puppet show, not the marionettist at the top controlling everything. "Constructive development does not assume a relatively simple mapping between genotype and phenotype, nor does it assign causal privilege to genes in individual development. Instead, the developmental system responds flexibly to internal and external inputs, most obviously through condition-dependent gene expression, but also through physical properties of cells and tissues and 'exploratory behaviour' among microtubular, neural, muscular and vascular systems" (6). Consequently, "Organisms are not built from genetic 'instructions' alone, but rather self-assemble using a broad variety of inter-dependent resources" (6). "'Reciprocal causation' captures the idea that developing organisms are not solely products, but are also causes, of evolution" (6). In a way, therefore, this is the flip side of constructive development, stressing the underlying causes and their modes of action. Unfortunately, ways of traditional thinking (SET) "typically fail to recognize that developmental processes can both initiate and co-direct evolutionary outcomes. By contrast, the EES views reciprocal causation to be a typical, perhaps even universal, feature of evolving and developing systems, characterizing both the developmental origin of phenotypic variation and its evolution in response to changeable features of its environment" (7). Helpfully, we are given a diagram to guide us (Figure 3.2).

Traditionally, development has been conceptualized as programed, unfolding according to rules and instructions specified within the genome. DNA is ascribed a special causal significance, and all other parts of the developing organism serve as "substrate" or "interpretative machinery" for the expression of genetic information. Evolutionarily relevant phenotypic novelty results solely from genetic mutations, which alter components of the genetic program. Under this perspective, organisms are built from the genome outward and upward, with each generation receiving the instruction on how to build a phenotype through the transmission of DNA. By contrast, in the EES, genes and genomes represent one of many resources that contribute to the developing phenotype. Causation flows both upward from lower levels of biological organization, such as DNA, and from higher levels downward, such as through tissue- and environment-specific gene regulation. Exploratory and selective processes are important sources of novel and evolutionarily significant phenotypic variation. Rather than containing a "program," the genome represents a component of the developmental system, shaped by evolution to sense and respond to relevant signals and to provide materials upon which cells can draw (7).

Drawing things together, the basic message of the New Biology (the EES) is clear and simple. Darwinism (SET) sees variation as random,

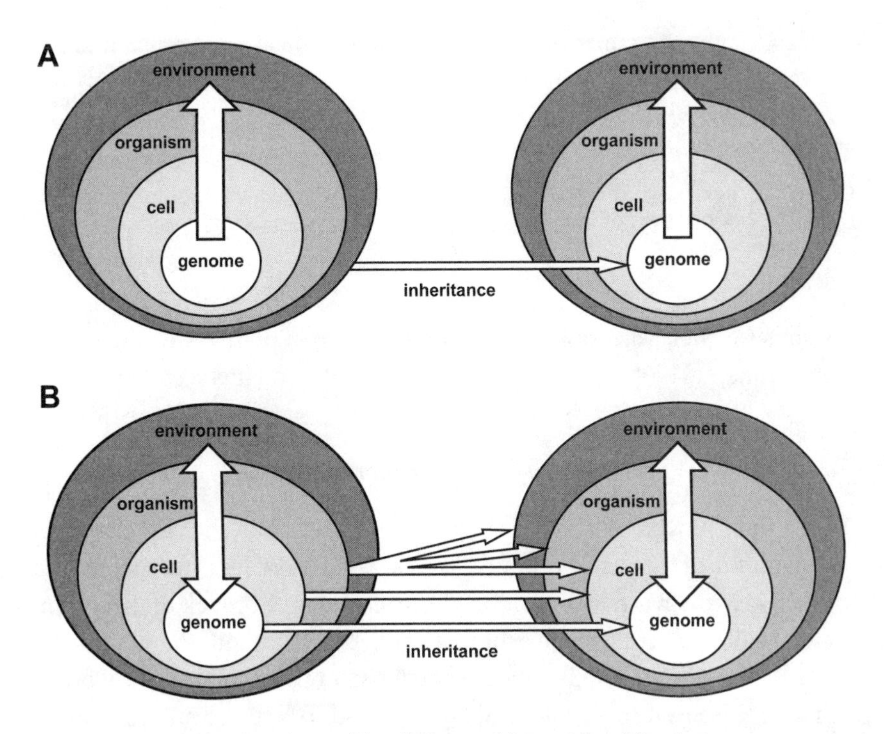

FIGURE 3.2 Contrasting views of development. (*a*) Programed development; (*b*) constructive development. *Credit:* Reformatted with permission of The Royal Society (UK) from Kevin Laland, Tobias Uller, Marcus W. Feldman, Kim Sterelny, Gerd B. Müller, Armin Moczek, Eva Jablonka, and John Olding-Smee, "The extended evolutionary synthesis: its structure, assumptions and predictions," *Proceedings of the Royal Society B,* 282:1813, 2015, Figure 1; permission conveyed through Copyright Clearance Center, Inc.

meaning nondirected, and all change comes from the external pressures of natural selection. This new approach sees forces as internal to the organism as important, if not more so, and this means / entails that variation is in some very important sense directed toward a desired or desirable end. Mechanism versus organicism.

> The EES is thus characterized by the central role of the organism in the evolutionary process, and by the view that the direction of evolution does not depend on selection alone, and need not start with mutation . . . As a consequence, the EES predicts that organisms will sometimes have the potential to develop well-integrated, functional variants when they encounter

> new conditions, which contrasts with the traditional assumption of no relationship between adaptive demand and the supply of phenotypic variation. (Laland et al. 2015, 8)

STANDARD EVOLUTIONARY THEORY

Let us take in order the three areas that the New Biologists highlight as reasons for their rejection of SET and embrace of the EES: evolutionary developmental biology, developmental plasticity, and niche construction.

Evolutionary Developmental Biology

In the light of what we have seen above, it is here that the real case is to be made for or against the EES. The rest is more or less frosting on the cake. Darwinism stresses randomness in the sense of lack of direction. Value is all relative. The New Biology puts direction into the variations, and in some sense, there is purpose and (absolutely) valued ends. In order to make the case for the Old Biology, we need to return again to history. Around 1959, there were gaps in the Darwinian theory (SET)—notably the "black box" of development taking us from genotype to phenotype—and, perhaps even more worrisome, a luddite attitude to molecular biology. We started this chapter with a few, entirely typical thoughts by Ernst Mayr on the topic. Needless to say, this is but part of the story. Go back to the 1930s, to Dobzhansky and the formulation of Neo-Darwinism or the synthesis. Dobzhansky worked with fruit flies—expectedly, since he had studied in the laboratory of Thomas Hunt Morgan, who had used fruit flies as he and his students worked out the physical nature of inheritance. Dobzhansky, however, did not stay in the laboratory. He went outside and studied fruit flies in the wild. And he came up with some interesting findings, notably that there are genetic differences between isolated populations of fruit flies as found in the mountains and valleys of the western part of the United States. Moreover, he came to see that natural selection plays a crucial role: "The data to be reported in the present article show that the genetic composition of populations in some localities changes with time. The changes appear to be connected with the annual climatic cycle and are consequently reversible. Nevertheless, since these changes demonstrate a great plasticity of the species genotype, they may be regarded as models of adaptive evolutionary changes on the racial and higher levels" (Dobzhansky 1943, 305).

One thing that was coming to impress Dobzhansky mightily was that there is much variation in natural populations, and whereas one might think that natural selection is always pushing populations toward genotypic similarity of its members—the "classical" view that there is one best form, and all others are or will be eliminated by selection—in fact, selection is promoting variation in populations, with different genotypes "balancing" out (Lewontin 1974). One obvious reason for such selection-promoted variation is selection for rareness. Suppose you have two forms, say red and green, and predators go after both but must learn that the forms are potential prey. Suppose that initially you have 90 percent red and 10 percent green, then the predator is much more likely to encounter red and to learn that that is potential prey. Green will be at a selective advantage. This means that proportionately green will start to do well, and its numbers will rise, first 20 percent, then 30 percent, and so on. Soon, it will no longer be the rare form, and the predators will learn that it is prey also. So now, green is harvested, and red less than it was, meaning that, other things being equal, one will get a balance of red and green in the population, brought on by natural selection.

The trouble was that no one could confirm or refute Dobzhansky's hypothesis, and it didn't help that Herman Muller, former student of Morgan and Nobel Prize winner, endorsed the classical view that selection works to eliminate variation. The fact that he knew little about variation in natural populations was irrelevant. He was a Nobel Prize winner and so had to be right by definition. Finally, in the early 1960s, the Gordian knot was cut by, of all people, Dobzhansky's former doctoral student, Richard Lewontin (1974). He with others (in the 1960s) showed that different-sized molecules in an electric field move through gels at different rates. One can thus sort out DNA strands indicative of different genes. This technique of "gel electrophoresis" was able to show that Dobzhansky wins hands down. Natural populations have lots of variation. Note, incidentally, that this brilliant work was about as reductionist as it is possible to get. Break things down to the smallest elements and get the answers. The irony will not be lost on the reader that it was done by Wilson-critic Richard Lewontin, of Spandrels fame (Ruse 1982). Yet more irony appears when one learns that the conference, at which Mayr made his above-quoted comments about the value of molecular biology, had in the audience none other than Mayr's former student, Richard Lewontin!

Lewontin's result was important for two reasons. First, *pace* Mayr, molecular biology could have and did have a vital role in the context of SET. Although note that, paradoxically, it was less part of the theory and more a handmaiden to solve already-existing biological problems. Second, and this is really important for us here, it removed at one stroke the big problem so many have, including—especially including—the New Biologists. It seems implausible to impossible that evolution can occur when, before selection gets to work, organisms have to hang around waiting for the right mutations to appear. Given they are random, this implies very long waits, and like the time and tide, evolution waits for no one. Climate change means that having a white coat is no longer effective camouflage. The background snow and ice are gone. Now, you have lush green pastures. To escape predators, you need a camouflaged (ideally, green, failing that some shade of brown) coat. As likely as not, if you wait for mutations, you will get a red / yellow / blue / black / white-again coat. Before the green / brown-coat-producing form of a gene appears, you will long be extinct.

This is why so many, including the EES supporters, want to put some guided direction into new variations. You need a green / brown-coat-producing variation. The EES can give it to you. Now, with Dobzhansky's balance position, you don't have to hang around waiting for the right variation to appear. Perhaps there already exists a variation in the population for green / brown-coat coloration. Or, if not that variation, other variations that will do the job of protecting you—something making you taste unpleasant to predators. They will soon start leaving you alone. Or making you nocturnal and so avoiding the daytime predators. Or with a tendency to hide—you are shy. Or perhaps a variation telling you to get the hell out of here and find a friendlier place to live! Of course, for didactic purposes, we are simplifying—a single gene is unlikely to make you nocturnal. But the central point is clear. You don't, at least in the eyes of

SET supporters, need directed mutations. Thanks to Bauplan-booster Richard Lewontin, you can see that what you have from random variations is good enough. Adaptive evolution through natural selection, although it is true that this reinforces the value-free nature of Darwinism, something within the machine metaphor. There is no absolutely right answer. What works works. There is no desired goal as the EES position asks for. The Darwinian is comfortable with this (Ruse 2006).

What about the "black box" situation? The SET supporter welcomes ideologically free evolutionary development: evo-devo (Arthur 2021). This is precisely how science—machine-metaphor science—works. But it is claimed that nothing turned up thus far in any sense challenges the mechanical approach of SET. Take by way of example cave fish. The Mexican tetra, *Astyanax mexicanus,* has invaded caves, and many forms, independently, have started to lose their sight. The eyes start off normally, but then they stop developing, and in some cases, this leads to total blindness. Adaptive reasons for this are easy to produce. Perhaps in the cave environment, eyes are of little use but rather are dangerous because the detritus in the cave could lead to irritations and infections. Or other such reasons.

However, blind cave fish have been seized on by EES enthusiasts as evidence of environmentally induced change that then gets incorporated in the gene line. Directed variation, meaning that you get a variation of use straight off without selection, produced by the organism itself. Lamarckism! Well-known enthusiast for the EES Eva Jablonka "thinks that heritable epigenetic changes alone could explain the loss of eyes. What is more, she even thinks it possible that the epigenetic changes were somehow triggered by the cave environment in the first place. That would be a form of Lamarckian evolution: the idea that characteristics acquired during an individual's lifetime can be passed on to descendants" (Le Page 2017). Expectedly, SET supporters will have none of this. "'This is a most interesting paper,' says evolutionary biologist Douglas Futuyma of Stony Brook University in New York. But he doesn't think it poses any challenge to standard evolutionary theory as the epigenetic change is itself most likely a result of a genetic change" (Le Page 2017).

What's going on here? Apparently, what is causing the eye loss is a phenomenon known as "methylation." This occurs when a set of molecules—a methyl group—attaches itself to other molecules, preventing or altering their function, such as a DNA molecule no longer being able to give out information—to take our case—on how to produce eyes. What excites someone such as Jablonka is that these methyl groups are activated by environmental conditions—to take our case—being in a cave with no light, and then apparently go backwards to the DNA molecule where they become part of it. And, of course, this can go on indefinitely. Inheritance of acquired characteristics! Not so fast, say SET supporters. Jerry Coyne (2017) writes of what is going on here:

> The thing is, this is not a violation of evolutionary theory in any sense. The commands for methylation under certain conditions, and the results of a gene being methylated, have evolved by changes in regular DNA sequences that control the methylation of *other* DNA sequences. It's just an evolved way to regulate genes.
>
> What has likely happened in these cave fish, then, is that other, "regulatory" genes have changed in the cave forms that provide instructions sort of like this: "hey, gene Y: at a certain point, you attach methyl groups to other genes for eye formation, shutting down those genes and causing the eyes to degenerate." That could easily evolve by conventional natural selection.

There are more pithy comments in Coyne (2019). We'll pause here in the midst of the debate to put things in context. There are going to be huge adaptive advantages in organisms being able to regulate their development according to environmental conditions. For instance, if an organism grows in an environment with limited food sources, there will be big adaptive advantages to growing up small with a correspondingly small appetite (Veenendaal, Painter, de Rooij, Bossuyt, Gluckman, Hanson, and Roseboom 2013). Conversely, in an environment with lots of food, growing up big with a large appetite may protect you from predators. From the perspective of natural selection, much more efficient for doing this sort of thing, rather than having to produce two sets of genes in two sets of organisms, where one set succeeds and the other set does not, is to have ways to regulate growth along the way, in one set of organisms. This is where methylation comes into play. It enables the organism to switch off genes and so forth, having ultimate phenotypic consequences. You don't need to start everything from scratch. But where do the instructions for methylation come from? Ultimately, the DNA. There are genes that tell the methyl groups to spring into action or not. And these controlling (regulating) genes are there because of natural selection. Unless they confer advantage—for instance, giving the organism the ability to vary its size and appetite adaptively—they are not going to be around for long.

Leave it at this. The reader can see how the arguments are framed and presented: how an EES enthusiast is going to make the case, and how the SET supporter will respond. The problem is that nature provides new challenges—let us say, a new predator or a change in climate. How is the organism to respond? For the EES supporter, in some sense, you need directed mutations. Changing color to avoid the predator. Growing a furry coat to protect against the cold. Not so fast says the SET supporter. Organisms, thanks to such phenomena as selection for rareness, always have a range of (genetically caused) variations to call on. No appropriate camouflage variations? Then how about a variation to move to a nocturnal existence? Or to become so disgusting in taste that the predator seeks elsewhere? Stay with mechanism. No need to move to organicism with its flavors of teleology. You have an end? Let nature supply it.

Developmental Plasticity

What of Waddington's contribution to the discussion about developmental plasticity? He promoted the idea of what he called "genetic assimilation," postulating a cause that involved canalization, arguing that "if an animal subjected to unusual environmental conditions develops some abnormal phenotype [such as having cross-veins in *Drosophila*] which is advantageous under those circumstances, selection will not merely increase the frequency with which this favorable result occurs, but will also tend to stabilize the formation of it, and the new development may become so strongly canalized that it continues to occur even when the environment returns to normal. For a series of events of this kind, the name 'genetic assimilation' may be suggested" (1953, 125). Somehow, it seems, in accordance with the organismic view, the organism itself has taken control of its fate. It is no longer going to be subject to the tyranny of natural selection.

Expectedly, SET supporters think little of this. Surely what is happening is that the cross-vein is controlled by several genes. Normally, taken individually, they have no effect on survival and reproductive success. When heat shocks are applied, this changes things sufficiently that they do have an effect. Most individuals, note, do not even now exhibit the cross-vein. Presumably, they don't even have one instance of a pertinent gene (or gene variation, allele). Selection on those that do show a cross-vein produces descendant organisms with several pertinent genes, and these, taken as a whole, have sufficient power to produce wings with cross-veins, even in the absence of heat shocks. SET supporters add (what Waddington acknowledges) that there is no reason to think that cross-veinedness is of adaptive advantage, so there is little reason to think there would be selection for it.

Today's SET supporters, such as Futuyma (2017), are prepared to allow that plasticity could in theory be important in evolution:

> Therefore, phenotypic plasticity could be said to truly play a leading role (with genes as followers) if an advantageous phenotype were to be triggered by an environment that really is novel for the species lineage, an environment that its recent ancestors did not experience and which, therefore, had not exerted natural selection. Of course, it is possible that a novel environment—a new pesticide, for example—could evoke a developmental effect that happens to improve fitness, just as it is possible that a random DNA mutation improves fitness.

He adds, more pessimistically: "But no theory leads us to expect such an effect to be especially likely." And this brings us to our third area of debate: niche construction.

Niche Construction

Richard Lewontin (1983) said that "organisms determine what is relevant" (280). As Douglas Futuyma (2017) remarks, this is blindingly obvious to any naturalist. "Even if a species does not literally construct its environment, like a beaver, it determines its environment by its behaviour and physiology. What is relevant to the life of an aerially foraging swift, to a foliage-gleaning warbler, and to a fish-eating loon (diver) is obviously very different" (4). And it is clearly going to be a two-way process. As the invader comes in, it will alter the original habitat, perhaps cleaning out certain species or fouling hitherto-clear waters. Conversely, the habitat will put selective pressures on the invader—hitherto-unencountered predators needing adaptive responses. Critics (such as Futuyma) of the New Biologists argue that niche construction "should be regarded, after natural selection, as a second major participant in evolution" and indeed as a "core evolutionary process" (4). The trouble is working out precisely what this means. It is obviously a core evolutionary process. But that has always been recognized:

> When we behold a wide, turf-covered expanse, we should remember that its smoothness, on which so much of its beauty depends, is mainly due to all the inequalities having been slowly levelled by worms. It is a marvelous reflection that the whole of the superficial mould over any such expanse has passed, and will again pass, every few years through the bodies of worms. The plough is one of the most ancient and most valuable of man's inventions; but long before he existed the land was in fact regularly ploughed, and still continues to be thus ploughed by earth-worms. It may be doubted whether there are many other animals which have played so important a part in the history of the world, as have these lowly organized creatures. (Darwin 1881, 313)

Charles Darwin on earthworms. The founder of natural selection theory. No comment.

RIVAL PARADIGMS: THE MACHINE METAPHOR

Clearly, what is at stake here is more than "just the facts." There are underlying metaphysical issues: a machine-based approach, SET, against an organism-based approach, the EES. The real question is why should the machine-based thinkers advocate their approach, and why should the organism-based thinkers reject this approach and opt for something different?

Starting with the machine-based thinkers, the mechanists, the supporters of SET, the answer about their choice is easy. Faced with something very similar to what Thomas Kuhn called different paradigms or metaphors, at once we have the consequence that appeal to the facts is never completely adequate. It is rather that the one paradigm opens up the way to "normal science" in a fashion that the other paradigm does not. It gives you problems to solve—actually, what Kuhn (1962) calls "puzzles," because puzzles have solutions, whereas problems may not—and work that can be completed. In a way, therefore, the answer is pragmatic. With mechanism, with SET, you can set about getting solutions in a way that is impossible with organicism, with the EES.

What would be a good example? We have seen how sociobiology was a tremendous extension of SET. The Darwin-based evolution of social behavior was left more or less untouched for a hundred years, and then it sprang into incredibly fertile life. Again, mechanistic—all going according to unbroken law—and reductionistic. Individual selection against group selection. Selfish genes rule, okay! Here is a beautiful example, work done by the English evolutionist Geoff Parker around 1970. The subject organism was the lowly dung fly, *Scatophaga stercoraria*. The males search for fresh cow pats. Females then arrive and mate with the males, after which they lay their eggs on the pats, and subsequently the hatched larvae bury down into the dung and use it for nutrition. Somewhat complicating this is the fact that the newly deposited pats are too runny for use, and a period of time is needed for a skin to form on the top of them. What makes all this an ideal subject for study is that there is intense sexual selection, for there are often about five males to every available female. This means that males must put in considerable effort if they are to mate successfully with a female. To this end, males position themselves on the field, spaced with respect to the position of the pats, trying to find the optimum place for finding available females. Too close to the fresh pats, and you don't find females; get a little further out, and the number of females rises sharply, but so too do the number of competing males. Further out, fewer females but also fewer competitors. "Males should be distributed between zones in such a way that all individuals experience equal expectations of gain. Hence the proportion of females captured in a given zone should equal the proportion of males searching there, assuming that all females arriving are equally valuable irrespective of where they are caught" (Parker 1978, 219–220). Parker found that the findings fitted the predictions very closely.

Once he has mated, the male's strategy changes. The job now is to get the mated female onto a skin-covered pat and to stay with her until she lays her eggs. A complicating factor is that with multiple matings, as sometimes happens, the last male to mate with the females is going to be the fertilizing male. So, males with females carrying their sperm have an interest in getting the females to oviposit before any other males can get at them. As the pats age, we expect to see a change in male strategy: earlier, out from the pats to where there are waiting females; later, into toward the pats where females can lay their eggs. Again, prediction and observation are very close.

Dung flies are the gift that keeps on giving. There are all sorts of testable subsidiary predictions. It takes half an hour to copulate and for the male to get all that sperm out in action. At some point, is it not better for the male to quit early and go and find a fresh mate, in the hope that you will be the final copulatory male? We will here leave these and related matters, referring the interested reader to the literature. Our point is made. The mechanical, reductionistic approach of SET to social behavior pays huge explanatory dividends, seeing a symmetry between prediction and explanation—once you have a successful prediction, you are in possession of an adequate explanation. Mechanism works. Pragmatism pays off!

RIVAL PARADIGMS: THE ORGANISM METAPHOR

Thus, the case for the mechanist. SET and the philosophy on which it rests are justified pragmatically. They lead to lots of good, new science. The EES supporter is hardly indifferent to this and, if challenged, is obviously going to claim something similar for their approach. But pull back for a moment and ask if one can look at things, if not entirely objectively, then a little less heatedly. For a start, for all the noisy fireworks being hurled around, in real life, there is going to be huge overlap between SET supporters and EES supporters. Neither side, for instance, is going to challenge the accepted picture of recent human evolution: out of the jungle and onto the plains, becoming bipedal and embracing the hunter-gatherer life, the explosion of brain power, the importance of sociality and the means to enact this, notably developing speech, and much more. For all that Kuhn talks about paradigms existing in different worlds, that is not really the case, at least not here. There will undoubtedly be different interpretations, but that is another matter. Just don't overestimate the differences. Moreover, don't overestimate the importance of such differences as there may be. If, for example, traditional Darwinians, SET supporters, can give an explanation of niche construction, so be it. It doesn't destroy the point that organisms can now be seen to control their destiny, as it were, and that is precisely the point of the organicist. If, to give another example, molecular biology can now explain development from a mechanistic perspective, this confirms organicism! It is now being seen that the organism itself, through its development, makes a significant contribution to ongoing evolution. It is not sola natural selection.

The whole point of the debate is that it is about different paradigms, different metaphors, different ways of looking at the world. The same facts can be interpreted in different ways. Most importantly, for the organicist, any world picture that does not make humans special is simply not doing its job. So, we know that mechanism cannot be the whole story—or if you prefer, the only reasonable story. More than this, whatever they say, mechanists recognize this too! They try to slide progress in by the back door by making it a consequence of natural selection. Go back to Julian Huxley and arms races. Requote him: "With life it has been the same: if one species happens to vary in the direction of greater independence, the inter-related equilibrium is upset, and cannot be restored until a number of competing species have either given way to the increased pressure and become extinct, or else have answered pressure with pressure" (Huxley 1912, 115). Adding: "So it comes to pass that the continuous change which is passing through the organic world appears as a succession of phases of equilibrium, each one on a higher average plane of independence than the one before, and each inevitably calling up and giving place to one still higher" (116). Mechanistic if you will, but making humans special in a way that strict Darwinians would not expect—or if they did expect, they would then ignore. Huxley incidentally was part of a tradition. In the third edition of the *Origin,* Charles Darwin floated the idea. Lines of organisms compete against each other, and their adaptations get ever more sophisticated:

> If we look at the differentiation and specialisation of the several organs of each being when adult (and this will include the advancement of the brain for intellectual purposes) as the best standard of highness of organisation, natural selection clearly leads towards highness; for all physiologists admit that the specialisation of organs, inasmuch as they perform in this state their functions better, is an advantage to each being; and hence the accumulation of variations tending towards specialisation is within the scope of natural selection. (Darwin 1861, 33)

And then, of all people, arch-reductionist Richard Dawkins is into the same game. "Directionalist common sense surely wins on the very long time scale: once there was only blue-green slime and now there are sharp-eyed metazoa" (Dawkins & Krebs 1979, 508). As always, it is the analogy with human progress that is the key:

> Computer evolution in human technology is enormously rapid and unmistakably progressive. It comes about through at least partly a kind of hardware / software coevolution. Advances in hardware are in step with advances in software. There is also software / software coevolution. Advances in software made possible not only improvements in short-term computational efficiency—although they certainly do that—they also make possible further advances in the evolution of the software. So the first point is just the sheer adaptedness of the advances of software make for efficient computing. The second point is the progressive thing. The advances of software, open the door—again, I wouldn't mind using the word "floodgates" in some instances—open the floodgates to further advances in software. (Ruse 1996, 469, quoting a conference presentation in Melbu, Norway, July 1989)

He adds, "I was trying to suggest, by my analogy of software / software coevolution, in brain evolution that these may have been advances that will come under the heading of the evolution of evolvability in the evolution of intelligence" (469).

In reaction, organicists will conclude that mechanists are into values as much as anyone. It is just that they pretend they are not. As we have noted, arms races are not quite the gift from heaven that people such as Darwin and Dawkins suppose them to be. Russia's invasion of Ukraine makes this very clear. You can have all the fancy weaponry you might imagine. This does not guarantee that you will get what you want. You can have all the computer power you might imagine. This does not guarantee the evolution of humans, certainly not humans as we recognize and desire. Arms races might lead to some very unwanted ends—the cannibalistic Morlocks of H. G. Wells's *The Time Machine,* for instance. For all that they seem to be a huge Darwinian success, they do not at once flood one with thoughts of progress. With good reason, organicists might conclude that someone is playing a three-card trick, slipping in absolute values when no one is looking. Remember the leading founder of Neo-Darwinism (SET), Theodosius Dobzhansky. It is he who gave the definitive mechanistic explanation of apparently guided variations. Natural populations are like a library. They always have a range of variations. You have to write an essay on dictators? If there is nothing in the library on Hitler, then perhaps there is something on Stalin. If not Stalin, then perhaps Mao. You have to have protection against a predator? If there is no color-camouflage variation, perhaps there is a variation for changed habit. If not habit, then perhaps foul taste. No need to start from scratch, trying to write a book on a dictator or providing the ideal antipredator variation. Yet, the same man who argued this was at the same time president of the US Teilhard de Chardin Society—the Jesuit priest who, as we shall see, was the organicist's organicist, offering up an idiosyncratic brew of Bergsonian vitalism and Catholic Christianity, a brew where human beings came bubbling to the top. The very existence of Theodosius Dobzhansky, a leading scientist of the twentieth century, suggests a little modesty and caution is recommended for both mechanists and organicists.

In the same spirit, warning against arrogance, there is another very important point. Given the triumph of mechanism, the culture of science is very strongly against absolute value judgments. Such judgments are for pop science—museums and television and magazines and the like. So, even if at a personal level you believe that nature has value and it shows increasing value as one gets near to humans, while at work as a professional scientist, you had better keep such thoughts to yourself. Otherwise, you risk saying goodbye to good university jobs, graduate students, funding, membership in the National Academy of Sciences, and the like (Ruse 1996). Organicists have very public values that guide their science.

Go back to the beginning of this chapter. People such as Julian Huxley and Edward O. Wilson are open in their belief in progress up to humankind, and they would think that a science that does not allow this is a science inadequate. We shall learn that the feminists and the Gaia enthusiasts are open in their belief that the dead matter of mechanism is the wedge for putting women in their place and demonstrating their supposed inferiority. A science that allows this is a science inadequate. The Aristotelians are open in their commitments to final cause thinking, believing it important not just epistemologically but ethically also. A science that does not allow this is a science inadequate. Yet, take heed, mechanists. Organicists may be completely wrong. Just don't let the vehemence of your convictions stem overly much from the comfortable, established status of your position. Think again about human evolution. Five million years as hunter-gatherers, living and surviving (and reproducing) in bands (tribes) of about fifty. What was the secret of our success? That we worked together as a team. In the immortal words of the metaphysical poet John Donne: "No man is an island." How did this come about? Through the most mechanistic of Darwinian processes. The already-mentioned reciprocal altruism: You scratch my back and I will scratch yours. Remember: "as the reasoning powers and foresight of the members [of a tribe] became improved, each man would soon learn from experience that if he aided his fellow-men, he would commonly receive aid in return" (Darwin 1871, 1, 163). Self aiding self. Probably reinforced by a good helping of kin selection: "we are evolutionarily primed to define 'kin' as those with whom we are familiar due to living and rearing arrangements. So genetically unrelated individuals can come to be understood as kin—and subsequently treated as such—if introduced into our network of frequent and intimate associations (for example, family) in an appropriate way" (Johnson 1986, 133). And yet, if one tried, could one think of a more holistic understanding? Humans throve because they worked together as a unit, as a whole. We are all in this together. Organicism! "Ask not for whom the bell tolls. It tolls for thee."

So, bringing this chapter to an end, we are not in a *Dragnet* situation: "Just the facts, ma'am. Just the facts." We have not so much a draw as a stalemate. Neither side can land a knockout blow, nor is it easy to see how they ever could. Values—absolute values—are all important. Mechanists want them out, and presumably they would claim that for such issues as progress, they are showing you can get the results with blind, value-free laws. Organicists want them in, and equally presumably, they would claim that without them, whatever the mechanist might say, you cannot get such things as the importance of humans without a value-laden paradigm or metaphor. In an interesting way, we have an echo of Kant in the *Critique of Judgment* on biology. We can explain organisms mechanically, but still there is a final-cause aspect to them. Kant's solution was to make final-cause thinking necessary but heuristic. For that reason, he judged biology second class. The organicist position is that you might be able to explain organisms mechanically, but (contrary to the claim of the mechanist) there is still the equivalent of final-cause thinking, namely values such as the significance of humans and the implications, such as treating all equally and providing a healthy world in which to live. Where the organicist differs from Kant is in denial of the belief that this dimension to understanding is merely heuristic and hence points to the second rate. Values are as important as facts, and even the mechanist acknowledges this implicitly. Hence, to get the full story, we must continue to explore values, and to do this, we must now go on into real life. Only then will we be in a position to draw the threads together.

Chapter Four

HEALTH

A wise man should consider that health is the greatest of human blessings, and learn how by his own thought to derive benefit from his illnesses.

—HIPPOCRATES

Up until now, we've explored organicism, mechanistic points of view, and how organisms change over time. In this chapter, we will take these debates one step farther: to "real world" examples in human health. Why? We (humanity, not specifically the two of us) know more about human health than the health of any other species, and more interestingly, there is almost certainly more to human health than to the health of other organisms. This may not be the case for physical health, but if we think of mental health, nonhumans probably haven't got much capacity either to feel the benefits of good mental health or to experience the harms that associate with poor mental health. True, some nonhumans can be bored or worse—think of the awful stereotypical behaviors that polar bears, used to roaming over vast areas, display if cooped up in too small a zoo—and it's almost universally agreed that at least mammals and birds and possibly some other animal taxa too can experience pleasure,[1] but

[1] Videos show examples of pleasure better than still photographs do. Try searching on YouTube for videos of cows let out onto pasture, having spent months over winter cooped up indoors. There's an example at https://www.youtube.com/watch?v=jQQTmuOEPLU.

there are presumably limits. Nonhumans, while they may get stressed, probably don't lie awake at night ruing a lost love or wake up sweating at the prospect of a job interview or worse.

We begin by considering four case studies—sickle cell disease, deafness, obesity, and schizophrenia—in each case seeing what reductionism and holism have to offer and whether there is any sort of middle way between a reductionist and a holistic approach to understanding human health. The issues of holism that we have discussed in earlier chapters (especially Chapter 2) apply with particular clarity when it comes to dealing with human health. Put crudely, when you get ill, do you want someone who finds out which bit of you is malfunctioning and tries to deal with that bit or do you want someone who sees you as a whole person and treats all of you not bits of you? There are arguments for and against each approach, and these two ways of understanding health embody different conceptions not only of health but also of biology in general. We close by considering by what is meant by "holistic medicine," an approach that eschews reductionism entirely.

SICKLE CELL DISEASE

Let's start with what at first sight seems a straightforward account of a human disease amenable to standard reductionist and mechanistic analysis: sickle cell disease, also known as sickle cell anemia. The usual account in textbooks goes something like as follows. Sickle cell disease is caused when someone inherits a faulty copy of the β-globin gene from both their parents. When this happens, the person throughout their lives suffers from what are called "sickle cell crises." These crises are caused by the person's red blood cells becoming more rigid and sickle-shaped, rather than their usual smooth biconcave shape. As a result, the cells are likely to stick together and block the passage of blood in the capillaries, preventing blood from reaching its destination. The affected red blood cells also don't live as long as normal cells, and so the person often develops anemia—hence sickle cell *anemia*.

The reason why sickle cell disease is relatively common, with about four to five million people suffering from it, is because individuals who have only one copy of the faulty gene (who are said to exhibit "sickle cell trait") experience almost no adverse effects but are less susceptible to malaria. In parts of the world where malaria is common, individuals with sickle cell trait are therefore fitter (in the sense that they end up having more children who themselves survive to reproduce) than individuals with sickle cell disease *and* than individuals who have two copies of the normal β-globin gene. This state of affairs, when individuals with one "normal" and one "faulty" copy of a gene are favored, is known as heterozygote advantage and is found in a number of other relatively common genetic diseases, for example cystic fibrosis, where heterozygotes are probably protected against tuberculosis (Mowat 2017). Most so-called genetic diseases, where there isn't something such as heterozygote advantage, are much rarer, as natural selection does its best (metaphorically speaking) to weed out these deleterious mutations.

So, at first reading, sickle cell disease provides a convincing example of reductionist biology—a mistake in the two copies of a person's β-globin gene (the one that they inherit from their mother and the one that they inherit from their father) results in the production of a faulty protein. This leads to a change in the appearance of the person's red blood cells, causing problems to the affected person. As a result, people with sickle cell anemia experience a range of symptoms, including the crises to which we have referred, which can be excruciatingly painful and frequently cause substantial internal damage. The net result is that people with sickle cell disease live on average about twenty years less than would otherwise be the case. But this story is a simplification. A fuller account is not only biologically more valid but more helpful for people with sickle cell disease.

For a start, sickle cell disease is best thought of not as a single disease but rather as a group of diseases. This is because there are a number of different mutations that can occur in the β-globin gene on chromosome 11. Then, there is the fact that the symptoms vary greatly from person to person. Some of this variation is due to which particular mutation(s) the person has, but some of it has other causes. The reality is that it is still not fully known why different people with sickle cell disease vary so much in the symptoms they experience. What is known, though, is that the worst symptoms, crises, can be triggered by a number of factors, including extremes of temperature, dehydration, lack of oxygen (such as being at high altitude, although it's OK to fly in commercial aircraft, as cabins are suitably pressurized), very strenuous exercise, infections, alcohol, smoking, and stress.

The reasons why such a wide range of factors can precipitate a sickle cell crisis are fairly well understood; for example, alcohol intake can lead to dehydration (alcohol is a diuretic; it reduces the body's production of antidiuretic hormone), and dehydration can set off a crisis, since lower water levels are more likely to cause the altered hemoglobin molecules that are made by the β-globin gene to clump together, a precursor to sickling. Low oxygen levels similarly seem to cause sickling by causing the altered hemoglobin to clump together as a result of changes in membrane permeability (Brugnara 2018).

Depending on the sort of person you are, this increasingly technical talk—"changes in membrane permeability" for example—may either cause your eyes to start to glaze over or lead you to think "at last we are getting on to what is actually causing a sickle cell crisis." However, we would caution against the hope that there is an "actual" cause. There are a number of causes; these causes operate at various levels, and suffering a symptom is not like striking a billiard ball and predicting where it will go—it's a lot more complicated for a host of very different reasons (and, anyway, what we know of modern physics means that we now realize that even the paths of billiard balls cannot be predicted with absolute accuracy). In saying that these causes operate at various levels, we mean:

- There are variations in the order of the nucleotide bases (adenine, cytosine, guanine, and thymine) that make up the DNA sequence of the β-globin gene; someone who has two particular variants of this sequence has the condition known as sickle cell anemia.
- The incidence of sickle cell anemia is relatively high worldwide, particularly in communities where malaria was present, because individuals with one particular variant of the DNA sequence of the β-globin gene are less likely to suffer from malaria and do not have sickle cell anemia.
- While we still do not know what triggers a particular sickle cell crisis, a wide range of factors can be involved—from ones such as extremes in temperature that can be characterized as "external" to ones such as stress that are the result of how the person concerned reacts to internal and / or external changes.

Furthermore, rather in the manner of Russian *(matryoshka)* dolls, investigating causes can be never-ending. We could write a whole chapter as to why individuals with one copy of the β-globin gene are less susceptible to malaria (cf. Luzzatto 2012). This plethora of causes—and these causes interact in a variety of ways—means that there are many ways of treating or attempting to prevent sickle cell anemia. The typical person with sickle cell anemia has about one crisis a year, a time of acute pain that lasts about a week or so and may require hospitalization; however, some people have an order of magnitude more crises and some get them only every few years. Pain management during crisis is a key part of treatment. Nonsteroidal anti-inflammatory drugs such as diclofenac can help, but many people require opioids (such as codeine, morphine, or their equivalents).

Crises are not only exceptionally painful, they can also be accompanied by substantial tissue damage. The most important thing therefore is to reduce the number of crises that people with sickle cell anemia experience. Traditionally, both children and adults were given folic acid to reduce the number of crises. However, the evidence that this works is not strong (Dixit et al. 2018; Williams et al. 2020). Until recently, the main thing that was advised was for people to avoid the various triggers we listed above. The first drug approved for the treatment of sickle cell anemia was hydroxyurea, which was shown in 1995 to decrease the number and severity of crises (Charache et al. 1995). By now, an increasing number of medicines have come onto the market, including a monoclonal antibody that binds to one of the proteins that plays a central role in the cellular process that can precipitate a crisis (Novartis 2019).

Other treatments include blood transfusions and bone marrow transplants, and there is a real possibility that gene editing may prove a game changer. Gene editing is already being tested, using the CRISPR (clustered regularly interspaced short palindromic repeats) technique (Hoban 2016), with thirty-three-year-old Victoria Gray becoming the first person to participate in the clinical trial. A year after she received the treatment, things are looking good. Prior to gene editing, Gray averaged seven hospitalizations a year; in the year since the treatment, she hasn't needed to be hospitalized once. As Gray put it: "I chose to participate in this trial because of hope—hope that it would change my life—and it has already in so many ways. I'm doing things for myself and for my family and just living life" (Plater 2020).

Gene editing is about as reductionist an approach to treatment as one could hope to find. If one does a literature search for holistic approaches to sickle cell disease, one finds some very different suggestions. Some of these draw on herbal approaches (e.g., Ameh et al. 2012), along with a component of holistic treatments. Some people presume that herbal treatments are unlikely to do much good, but the reality is that many of the medicines we use today come from plants (e.g., aspirin, digoxin, opiates, quinine, taxol, and chemotherapeutic agents from the Madagascar periwinkle) or fungi (e.g., myriocin and penicillin) and often had a long history of use by indigenous peoples or in folk medicine before being amalgamated into conventional medicine. Other holistic approaches to sickle cell disease include a seven-year study of the effects of preventive health and nutritional education, prompt treatment of illness and free supplies of vitamin supplements, malarial prophylactic and other necessary medication (Akinyanju et al. 2005), and an approach called SICKLE—which stands for Skin assessment, Infection control, Compression, Keep moving keep debriding, Local strapping and Endless support (Rivolo 2018).

A final point. If you consult the UK National Health Service (NHS) website for sickle cell disease, the last thing it advocates under the heading "Self-help for Treating a Sickle Cell Crisis" is "distractions to take your mind off the pain—for example, children might like to read a story, watch a film or play their favourite computer game" (NHS 2019a). This may sound like a council of despair (a bit like helpful tips before you undertake surgery on yourself—"sterilize your penknife . . ."), but there is an important truth here that is relevant to our overall argument. Sickle cell pain is the result of damage to the body, but as with all pain, it is also a product of the human mind. The human mind has a remarkable capacity to vary the extent to which we feel pain, depending on our mood, what we are thinking of, the behavior of others around us, and so on.

Sickle cell disease illustrates the shortcomings of approaches that focus only on mechanism / reductionism *and* of approaches that focus only on organicism / holism. Our next case study shows the importance of the social context within which the condition occurs.

DEAFNESS

Deafness, hearing impairment, or hearing loss all refer to the total or partial inability to hear sounds that most people can hear. Deafness is usually distinguished from hearing impairment or loss by defining it as an inability to hear speech even when it is amplified. Profound deafness (also sometimes referred to as being "totally deaf") means that a person is unable to detect any sounds (at any rate, any sounds below 90 dB—the sort of level of noise that requires one to shout to hold a conversation).

An important distinction can be made between people who are deaf and people who are Deaf. As with many distinctions, not everyone would agree with every aspect of the difference between the two, but the fundamentals are widely agreed. There are two main understandings of the word "Deaf." First, Deaf people have been deaf all their lives or, at any rate, since before they started talking or soon after. Second, and perhaps more importantly, Deaf people tend to communicate in sign language as their first language and see themselves as members of the Deaf community.

Issues around deafness and Deafness have been explored in fiction in a growing number of films—from *Children of a Lesser God* in 1986 to *Sound of Metal* in 2019. An example of an actual Deaf community in action is provided by Gallaudet University, Washington, DC. As it says on its website:

> Gallaudet University, the world's only university in which all programs and services are specifically designed to accommodate deaf and hard of hearing students, was founded in 1864 by an Act of Congress (its Charter), which was signed into law by President Abraham Lincoln. (Gallaudet University 2021)

Oliver Sacks, as always, puts it very well:

> When I had visited Gallaudet in 1986 and 1987, I found it an astonishing and moving experience. I had never before seen an entire community of the deaf, nor had I quite realized (even though I knew this theoretically) that Sign might indeed be a complete language—a language equally suitable for making love or speeches, for flirtation or mathematics. I had to see philosophy and chemistry classes in Sign; I had to see the absolutely silent mathematics department at work; to see deaf bards, Sign poetry, on the

> campus and the range and depth of the Gallaudet theater; I had to see the wonderful social scene in the student bar, with hands flying in all directions as a hundred separate conversations proceeded—I had to see all this for myself before I could be moved from my previous "medical" view of deafness (as a condition, a deficit, that had to be "treated") to a "cultural" view of the deaf as forming a community with a complete language and culture of its own. (Sacks 1989)

The distinction that Sacks draws between a "medical" and a "cultural" view of deafness is one that has grown in importance in disability studies. In the social model of disability, unlike the medical view, impairment is not denied but is not seen as the cause of disabled people's economic and social disadvantage (Oliver & Barnes 2010). Indeed, impairment can be distinguished from disability in that impairment is individual and private, whereas disability is structural and public (Shakespeare 2006). In the social model, it is society that needs to change, not disabled people who need to accept limitations. Of course, this shift in perspective doesn't solve everything, but it can be transformative.

Perhaps nowhere is this clearer than with respect to the Deaf community. Here is a quotation from Caroline O'Neill, born in 1980 and deaf since the age of five as a result of meningitis:

> I have been asked on many occasions whether I would take the opportunity to regain my hearing. My automatic answer is always, "No, of course not!" Being honest with myself, would I really say that if the technology was available? Answering a hypothetical question is one thing, but in real life? The technology isn't there yet, but what will happen when it is available? I love being Deaf, it makes me the wacky, opinionated individual I am today, and I wouldn't change that (bar wanting to lose a few pounds, but that's beside the point!) Yet . . . I would like to know what it is like to listen to music; or what it is like to walk down the street talking on a mobile phone. Surely everyone has wondered at some point? (O'Neill 2003)

At the time when O'Neill was writing, the technology to enable profoundly deaf/Deaf people to hear was indeed rare; now, it is more common. Best known and most widely used are cochlear implants, a mechanistic approach to tackling deafness. These are small, electronic devices that have both an external part (consisting of a microphone and associated components) and an internal part, which is principally a series of electrodes that carries impulses directly to different regions of the auditory nerve (National Institute on Deafness and Other Communication Disorders 2017). The net result is not that hearing is "restored" but rather that a person can learn to identify the various stimuli to their auditory nerve as meaningful sounds. In the United States, the first cochlear transplants were only used by adults, but in 2000, the US Food and Drug Administration approved them for use in eligible children from the age of twelve months.

Cochlear implants are not perfect, and as with any surgery, there is always a small risk that something could go wrong. Nevertheless, they are generally very successful. In one study that followed up 132 children ten years after they had received such implants, the implants enabled the children to manifest "a hearing performance and an appropriate language acquisition, currently comparable to normal hearing children" (Peixoto et al. 2013, 462). If one searches the Internet, one can find loads of videos with children and their parents saying how wonderful their cochlear implant has been for them.

But these "inspirational" videos divide viewers. As Lilit Marcus, a Child of Deaf Adults (CODA) and involved in the Deaf activist community, puts it: "the 'deaf person hears for the first time' videos don't make me smile. They make me want to throw my computer out a window" (Marcus 2014). As Cooper concludes:

> The most erroneous message the videos propagate is that cochlear implants fully transform deaf individuals into hearing ones. With present technology cochlear implants are a tool, not a cure. The most successful cochlear surgeries never restore full, natural hearing. Many recipients struggle to distinguish sounds, particularly in environments with a lot of background noise. (Cooper 2019, 470)

But there is something much deeper going on here than the impression one gets from watching videos about cochlear implants. It surprises many hearing people to learn that there is controversy about the worth of cochlear transplants in particular and so-called treatments for deafness in general. Within the Deaf community, the very notion of such treatments is generally regarded with deep suspicion. For many Deaf people, being Deaf is an identity to be celebrated not a disability to be cured. Furthermore, the increasing frequency of cochlear implants reduces the size of the Deaf community so that cochlear implants have been likened to genocide. Indeed, one of the signs for cochlear implant is a two-fingered stab to the back of the neck, indicating a "vampire" in the cochlea (Cooper 2019).

This is not a bioethics textbook, so we will not pursue the question as to whether Deaf parents have the right to prevent their children from receiving cochlear implants or even whether they have the right to use assisted reproductive technologies, such as IVF, to ensure that their children are born deaf—something that the British couple, Tomato Lichy and Paula Garfield, who already had a deaf child, wanted to do, so that their second child would also be deaf (Appel 2019). Rather, the significance for our purposes of this movement within the Deaf community is in how it troubles conventional notions of health. Most of us simply assume that it is better to be able to hear than to be deaf.

This controversy can best be understood by a more holistic conception of health in which relationships can matter more than physical abilities. Imagine you and your partner are Deaf and have decided, for whatever reason, not to have cochlear implants—remembering that cochlear implants are not a panacea and do not work for everyone. If your deaf/Deaf children have cochlear transplants, chances are they will never be as fluent at sign language as they would have been. They won't be able to communicate with you, nor you with them, as would have been the case had they remained as they had been from birth. Their friends will include people with whom you may be unable to communicate. Not only is it not surprising that you might want your children to be Deaf, you might react to the suggestion that they receive cochlear implants as many parents who speak minority languages have reacted when their children were prevented from speaking such languages in school and forced to learn the dominant language—a standard tool of colonial powers. You might sincerely believe that it would be better for your children, healthier for them, to be part of the Deaf community.

Not many of us are deaf, but many of us are overweight. We now consider what reductionist and holistic approaches to obesity have to offer, with regards both to understanding its causes and to addressing the issue.

OBESITY

Many people are not content with their weight—both authors would be very happy painlessly to lose some weight, if it then magically stayed off. How happy people are with their size and appearance varies from country to country. An international online survey in 2015 with just over 22,000 adults from twenty-five countries found that people in Indonesia, Saudi Arabia, and Oman are most likely to say they are happy with their weight and body image; those in Hong Kong, Chile, and France are least likely. Women are generally less happy than men—but in some countries, such as Saudi Arabia, the opposite is the case—and in most countries, older adults are happier than younger ones, a notable exception being France where 68 percent of eighteen- to twenty-four-year-olds said that they were happy with their looks compared to just over 55 percent of those aged fifty-five or more (Gammon 2015).

These data are about people's subjective perceptions. What about more objective data on weight and health? The mantra to which many of us have been exposed is that our body mass index (BMI) should lie between 18.5 and 25 kg/m^2. Below that, we are underweight; above that, we are overweight. Indeed, above 30 kg/m^2, we are deemed to be obese. If your BMI is above 35 kg/m^2 and you have obesity-related health conditions such as high blood pressure or diabetes, you are said to be morbidly obese, as you are if your BMI is above 40 kg/m^2 and you have no obesity-related health conditions. It's a bit fiddly to calculate one's BMI—it equals a person's mass in kilograms divided by the square of their height in meters—but there are plenty of Internet sites that do if for you whether you use metric or imperial units.

In the history of humankind, being underweight has been far more of a problem than being overweight, but that is no longer the case. One very large study (on 19.2 million adults from 186 countries) concluded that in 1975, 13.8 percent of men were underweight, and 3.2 percent were obese;

by 2014, the corresponding figures were 8.8 percent and 10.8 percent. In 1975, 14.6 percent of women were underweight, and 6.4 percent were obese; by 2014, the corresponding figures were 9.7 percent and 14.9 percent (NCD Risk Factor Collaboration 2016). Age-standardized mean BMI in men increased from 21.7 kg/m^2 in 1975 to 24.2 kg/m^2 in 2014, and in women from 22.1 kg/m^2 in 1975 to 24.4 kg/m^2 in 2014. This means that globally, the average adult, for a given age, became about 1.5 kg heavier each decade over those four decades. The largest increase in men's mean BMI occurred in high-income English-speaking countries and in women's in central Latin America.

Now, BMIs aren't a perfect way of determining whether someone is obese (readers are advised to consult their physicians rather than drawing conclusions about their own health from anything we write). The advantage of BMI is that it's easy to determine and has been around for a long time (since the mid-1800s), so that it's widely used, in contrast to direct measures of obesity such as underwater weighing (densitometry) and DEXA (dual-energy X-ray absorptiometry) scans. However, BMIs provide indirect measures of obesity and don't take age, sex, bone structure, fat distribution, or muscle mass into consideration (Rothman 2008). For example, for most people as they age, muscle mass decreases but body fat increases—if these two cancel each other out, BMI will remain constant, and we may kid ourselves that we aren't getting any fatter when we are. Then, there is the opposite issue that some people who are very fit have BMIs that suggest they are overweight when, in reality, this is a result of more muscle not more fat (the authors are confident this is the case for them . . .).

Most of us are familiar with a litany of adverse health consequences of being overweight. The Centers for Disease Control and Prevention (CDC; within the US Department of Health and Human Services) list them as: all causes of death (which is pretty comprehensive); high blood pressure; high LDL cholesterol, low HDL cholesterol, or high levels of triglycerides; type 2 diabetes; coronary heart disease; stroke; gallbladder disease; osteoarthritis; sleep apnea and breathing problems; many types of cancer; low quality of life; mental illness such as clinical depression, anxiety, and other mental disorders; body pain and difficulty with physical functioning (Centers for Disease Control and Prevention 2021).

So, without in anyway intending to minimize the problems of being underweight, our focus is on the causes of these big historical rises in obesity and what might be done to tackle obesity.

Causes of Obesity

In one sense, the cause of obesity is obvious: it results from a long period of time when one's daily energy intake exceeds one's energy expenditure to the extent that one puts on a lot of weight (this needs a bit of finessing for growing children and women in the latter stages of pregnancy, but the principle is clear). But what causes this?

From an evolutionary perspective, the three big problems for all animals (including humans) are finding the right sort of food, avoiding being eaten, and (for those that reproduce sexually) finding a mate with whom to produce viable offspring. Throughout human evolutionary history, the problem with food has generally not been eating too much but eating too little—as the above data on recent historical changes in how many people are underweight or overweight show. Furthermore, we evolved well before we could rely on regular meals. Far better, therefore, to stuff ourselves when food is plenty in preparation for times when it is scarce. It's just that nowadays, for many of us—never forgetting the many people who do go hungry in even the most "developed" nations in the word—it's rare for us ever to get to the stage of being short of food for extended periods of time.

William Howard Taft (1857–1930) was the twenty-seventh president of the United States and is generally agreed to have been the biggest US president of all time. He was president from 1909–1913, stood 5 ft. 11.5 in. tall and his weight peaked at 335–340 lb. toward the end of his presidency. That's a BMI of 46 kg/m^2. It is perhaps unsurprising that he probably suffered from obstructive sleep apnea (Sotos 2003). Nowadays, even an overweight president receives plenty of comments about his weight. It is virtually inconceivable that a contemporary president could reach a BMI of 46 kg/m^2. Joe Biden would have to weigh in at 24 stones, that's 336 lb., to have a BMI of 46 kg/m^2; in fact, on his seventy-ninth birthday, on November 20, 2021, he was reported to have a BMI of 25 kg/m^2 (Zitser 2021). In Taft's time, size, in a man, was more likely to be associated with professional success and affluence. Going back to prehistory, large size in women may also have been highly regarded, perhaps because of a recognition that underweight women were less likely to reproduce successfully. The Venus (Woman) of Willendorf, dating from about 30,000 years ago, is one of a number of many so-called Venus figurines that survive from Paleolithic Europe at a time when Europe was in the grip of a severe ice age (Dixson & Dixson 2011). By today's standards, she looks overweight.[1]

It is easy to blame individuals who are obese for their body weight. And it is easy for obese individuals to blame themselves. However, most contemporary understandings of obesity use a model derived from the "social determinants of health" approach, which has a "layered" approach to understanding our health (Figure 4.1). In this model (Dahlgren & Whitehead 1991), the outermost layer consists of the general socioeconomic, cultural, and environmental conditions, such as the availability of work; the next layer consists of living and working conditions, including such things as education, sanitation, and housing; the next layer consists of social and community networks, such as family and neighbors; the next layer consists of individual lifestyle factors, such as physical activity and alcohol use; finally, we come to personal characteristics, such as age, sex, ethnicity, and hereditary factors.

It is clear that the Dahlgren and Whitehead model for health determination shown in Figure 4.1 is holistic. Indeed, if one thinks about it, the general pattern of weight increase in recent decades that we referred to above shows that a suitable understanding must focus on what is going on at several layers—any attempt to explain global trends in obesity that focuses only on what individuals are choosing to do is missing out on other factors.

Nevertheless, the holistic approach of Dahlgren and Whitehead is still quite rare in respect of the causes of obesity. Here, in order, are the nine most common causes of obesity as listed by the popular medical website

[1] We included a photograph of the Venus of Willendorf with our manuscript, but one prepublication reviewer wrote, "I would recommend eliminating the image of the Venus of Willendorf. This is just too gendered and it might offend some people."

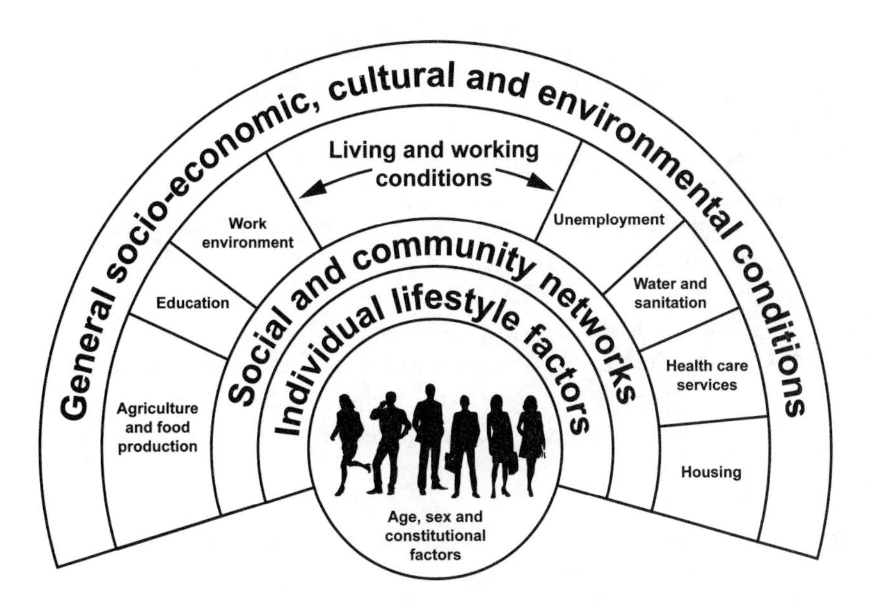

FIGURE 4.1 The social determinants of health model of Dahlgren and Whitehead. *Credit:* Reprinted with permission from Gören Dahlgren and Margaret Whitehead, "Policies and strategies to promote social equity in health," Figure 1 (Stockholm, Sweden: Institute for Futures Studies, 1991).

MedicineNet (2021): physical inactivity, overeating, genetics, a diet high in simple carbohydrates, frequency of eating, medications, psychological factors, diseases, and social issues. Only the last of these gets to a level above the individual, where the following is written: "There is a link between social issues and obesity. Lack of money to purchase healthy foods or lack of safe places to walk or exercise can increase the risk of obesity" (MedicineNet 2021). When one goes to what is currently ranked as the top-ranking site for health in the world, Healthline.com (SimilarWeb 2021)—ahead of the National Institutes of Health, the CDC, and the NHS—the entire entry on "What causes obesity?" is as follows:

> Eating more calories than you burn in daily activity and exercise—on a long-term basis—can lead to obesity. Over time, these extra calories add up and cause weight gain.

> But it's not always just about calories in and calories out, or having a sedentary lifestyle. While those are indeed causes of obesity, some causes you can't control.
>
> Common specific causes of obesity include:
>
> - genetics, which can affect how your body processes food into energy and how fat is stored
> - growing older, which can lead to less muscle mass and a slower metabolic rate, making it easier to gain weight
> - not sleeping enough, which can lead to hormonal changes that make you feel hungrier and crave certain high-calorie foods
> - pregnancy, as weight gained during pregnancy may be difficult to lose and might eventually lead to obesity
>
> Certain health conditions can also lead to weight gain, which may lead to obesity. These include:
>
> - polycystic ovary syndrome (PCOS), a condition that causes an imbalance of female reproductive hormones
> - Prader-Willi syndrome, a rare condition present at birth that causes excessive hunger
> - Cushing syndrome, a condition caused by having high cortisol levels (the stress hormone) in your system
> - hypothyroidism (underactive thyroid), a condition in which the thyroid gland doesn't produce enough of certain important hormones
> - osteoarthritis (OA) and other conditions that cause pain that may lead to reduced activity. (Healthline.com 2021)

Again, if one compares this list with the approach of Dahlgren and Whitehead in Figure 4.1, Healthline.com's list is all at the two inner levels of personal characteristics and individual lifestyle factors; there is no mention of the outer three levels: social and community networks; living and working conditions; or general socioeconomic, cultural, and environmental conditions. This almost exclusive emphasis on health policies targeted at getting individuals to change their behaviors is despite the fact that such an approach is rejected by academics in the field:

> . . . a growing body of research shows these types of health policies don't work. This is because such policies place responsibility on the person, ignoring the other drivers of obesity. Social inequity, the influence of food and beverage industries, and specific aspects of globalisation (including trade liberalisation) are all known causes of obesity. (Godziewski 2020)

Tackling Obesity

So, given that there are intertwined and multiple causes of obesity, even if this isn't always recognized (Ulijaszek & McLennan 2016), how might obesity be tackled? If one looks, in the UK, at the current official NHS advice, one reads the following:

> **TREATING OBESITY**
>
> The best way to treat obesity is to eat a healthy, reduced-calorie diet and exercise regularly. To do this you should:
>
> - eat a balanced, calorie-controlled diet as recommended by your GP [general practitioner] or weight loss management health professional (such as a dietitian)
> - join a local weight loss group
> - take up activities such as fast walking, jogging, swimming or tennis for 150 to 300 minutes (two-and-a-half to five hours) a week
> - eat slowly and avoid situations where you know you could be tempted to overeat
>
> You may also benefit from receiving psychological support from a trained healthcare professional to help change the way you think about food and eating.
>
> If lifestyle changes alone don't help you lose weight, a medication called orlistat may be recommended. If taken correctly, this medication works by reducing the amount of fat you absorb during digestion. Your GP will know whether orlistat is suitable for you.
>
> In rare cases, weight loss surgery may be recommended. (NHS Inform 2021)

This is not an approach that sees obesity as resulting from causes operating at a range of levels. The onus is put squarely on individuals and the need for individuals to change. Nor should it be thought that this individualistic approach is being advocated simply because it is found on a health service targeted at individuals who are searching online. Many of the most widely used models of behavioral change, both specifically for obesity and more generally, have a similar approach. The Capability, Opportunity, and Motivation Behaviour (COM-B) model of behavior change "recognises that for any behaviour to occur, the individuals concerned must have the physical and psychological capability to enact it, the physical and social opportunity, and be motivated to do it more than any potentially competing behaviour on relevant occasions" (Atkins & Michie 2013, 30). Here, there is mention of "social opportunity," but the onus is clearly on individuals.

A very different approach to tackling obesity is sometimes taken. Here is part of the Executive Summary of a report on obesity produced in 2007 by the UK Government's Foresight Programme:

> People in the UK today don't have less willpower and are not more gluttonous than previous generations. Nor is their biology significantly different to that of their forefathers. Society, however, has radically altered over the past five decades, with major changes in work patterns, transport, food production and food sales. These changes have exposed an underlying biological tendency, possessed by many people, to both put on weight and retain it. (Foresight 2007, 5)

The Foresight report uses the term "obesogenic environment" to refer to the role environmental factors may play in determining both energy intake and expenditure (cf. Noonan 2020). Not everyone may warm to the language of "obesogenic environments," but to give an obvious example, the prevalence of cars, computers, and TVs (part of our environment) means that most of us spend substantially more time sitting down than our grandparents did, which reduces energy expenditure and so is obesogenic (tends to cause obesity). There is now a large body of public health literature that explores the relationships between the built environment, physical activity, and obesity (Jamme et al. 2018). Although high-quality data are hard to come by, it is generally agreed that children in most countries are less likely nowadays to engage in unsupervised play an appreciable distance from home (Nordbakke 2019). The Foresight report points out that we walk less than we used to, and the proportion of us who have jobs that require substantial physical effort has also declined. Our eating habits too have changed very considerably. Food is cheaper, relative to average incomes, than it ever has been (notwithstanding occasional hikes in food prices, such as that in 2022), but this reduction in price has been greatest for energy-dense and nutrient-poor foods (fats, oils, starches, and sugars), whereas fruits and vegetables now cost more, relative to other food items.

As the Foresight report concluded:

- The causes of obesity are complex and multifaceted, pointing to a range of different solutions.
- At the heart of this issue lies a homeostatic biological system that struggles to maintain an appropriate energy balance and therefore body weight. This system is not well adapted to a changing world, where the pace of technological progress and lifestyle change has outstripped that of human evolution.
- Human biology, growth and development early in life, eating and physical activity behaviours, people's beliefs and attitudes and broader economic and social drivers all have a role to play in determining obesity. (Foresight 2007, 59)

It is perhaps unsurprising that so far, few interventions to reduce obesity at a population level have had great success. Nevertheless, there have been some promising results for interventions to reduce childhood obesity and these have entailed population-wide and / or community-based initiatives (World Health Organization 2012). Approaches include restrictions on the marketing of unhealthy foods and nonalcoholic beverages to children, nutrition labeling, food taxes and subsidies, policies that reduce barriers to physical activity (such as transport policies, policies to increase space for recreational activity, and school-based physical activity and food policies), strong community engagement at all stages of the process, careful planning of interventions to incorporate local information, and integration of the programs into other initiatives in the community.

All this takes time, as does changing societal attitudes. It is worth remembering how long it has taken, and is still taking, to reduce the incidence of cigarette smoking in the West (though its incidence is still increasing in some parts of the world) and how this reduction has come about by adopting a systematic approach, looking at all levels of the ecosystem (including restrictions on where cigarettes may be sold and on their advertising, and prohibitive taxes), rather than only telling individuals to stop smoking. As an anecdote, and to illustrate the change in public opinion, when the UK's BBC TV produced a widely acclaimed three-part drama in 2020 about serial killer Dennis Nilsen, who murdered at least a dozen young men and boys, kept their bodies in his home for a number of weeks, and then dissected and disposed of them by burning or flushing down a lavatory, there were hundreds of complaints. But these were not to do with the murders or their portrayal, but rather about the characters who were shown smoking cigarettes (P. Thomas 2020).

A lot is known about the causes of obesity; much less is known about the causes of schizophrenia, a mental illness and our final case study in this chapter. However, approaches to its treatment illustrate both the advantages and the disadvantages of adopting only a mechanistic / reductionist approach or only a holistic / organismic approach.

SCHIZOPHRENIA

The human brain is very possibly the most complicated structure in the universe. Each of us has one, with a typical mass, for an adult, of about 1,200–1,400 g (3 lb.). People with psychosis—a condition that can be temporary or lifelong—have difficulty distinguishing what is real from what is not real. There are many types of psychosis; for example, between 0.1 percent and 0.3 percent of mothers experience postpartum psychosis (VanderKruik et al. 2017), although with treatment, this nearly always clears up within about six to ten weeks. Psychosis can also result from sleep deprivation, certain medicines, and certain other drugs. In the

United States, about 3 percent of people experience psychosis at some point (National Institute of Mental Health 2021).

Schizophrenia is a mental illness characterized by periods of psychosis. An example of someone with schizophrenia is provided by August Natterer (1868–1933). On April 1, 1907, he had a pivotal hallucination of the Last Judgment:

> I saw a white spot in the clouds absolutely close—all the clouds paused—then the white spot departed and stood all the time like a board in the sky. On the same board or the screen or stage now images as quick as a flash followed each other, about 10,000 in half an hour . . . God himself occurred, the witch, who created the world—in between worldly visions: images of war, continents, memorials, castles, beautiful castles, just the glory of the world—but all of this to see in supernal images. They were at least twenty meters big, clear to observe, almost without color like photographs . . . The images were epiphanies of the Last Judgment. Christ couldn't fulfill the salvation because he was crucified early . . . God revealed them to me to accomplish the salvation. (UK Disability History Month 2017)

This vision precipitated an unsuccessful suicide attempt. Natterer thereafter maintained that he was "Redeemer of the World" and the illegitimate child of Napoleon Bonaparte. He was diagnosed with schizophrenia and spent the remaining twenty-six years of his life in a series of mental asylums, where he spent much of his time drawing the 10,000 images that he believed he had seen in his original hallucination. For posterity, we are fortunate that Natterer was visited by the pioneering psychiatrist and art historian Hans Prinzhorn, who brought his work and that of others to public attention through his writing. Prinzhorn amassed a collection of more than 20,000 pieces of artwork, some of which were exhibited in Hitler's notorious *Entartete Kunst* (Degenerate Art) exhibition in 1937, the most visited art exhibition of all time, four years after Prinzhorn had died.

Schizophrenia is relatively rare, with an incidence of about three cases per 1,000 individuals, but that still means that about twenty million people worldwide have schizophrenia. It has a diversity of symptoms: hallucinations (hearing, seeing, or feeling things that are not there); delusions (fixed false beliefs or suspicions not shared by others in the person's culture and that are firmly held, even when there is evidence to the contrary); abnormal behaviors (wandering aimlessly, mumbling or laughing to oneself, strange or unkempt appearance); and disorganized speech and/or disturbances of emotions (marked apathy or disconnect between reported emotion and what is observed such as facial expression or body language) (World Health Organization 2019a).

Tackling Schizophrenia

Most mental health and medical organizations and many experts in the field agree that it's premature to speak of identified "causes" of schizophrenia. As the UK's NHS puts it: "The exact causes of schizophrenia are unknown. Research suggests a combination of physical, genetic, psychological and environmental factors can make a person more likely to develop the condition" (NHS 2019b). Thomas Insel, in a highly cited article entitled "Rethinking Schizophrenia," noted that "After a century of studying schizophrenia, the cause of the disorder remains unknown" (Insel 2010, 187). It is clear that certain people are more at risk of schizophrenia—there is a genetic component, and birth complications (premature labor, a lack of oxygen during birth) increase one's chances of developing schizophrenia. Changes in the balance between the neurotransmitters dopamine and serotonin can trigger schizophrenia, as can adverse childhood experiences, stress, and drug misuse (especially cannabis) (Vallejos et al. 2017; NHS 2019b).

So, how does one tackle something when its causes are obscure? A mechanistic approach is rather ruled out, as we know so little of the mechanisms involved. We can begin by noting that schizophrenia needs tackling. Recovery rates are low, and people with schizophrenia are much more likely to be unemployed and to end up homeless or in the criminal justice system (Insel 2010). In addition, between about 5 percent and 20 percent of homicides are carried out by people with schizophrenia (Rund 2018).

In the late nineteenth and early twentieth century, schizophrenia was called dementia praecox ("premature dementia" or "precocious madness"), and attempts were made to treat it by hypnosis, psychoanalysis, occupational activities, drugs such as opiates and barbiturates (to relieve distress), and injections of glandular extracts. The attempts by psychoanalysts to treat schizophrenia subsequently became seen as providing a cautionary tale (Willick 2001). Psychoanalysts were convinced that schizophrenia resulted from "damage to the ego and its functions [which] took place primarily in the first one or two years of life" and "that the major cause of these ego impairments was to be found in profoundly inadequate caretaking" (Willick 2001, 28). Willick went on to write: "The neurobiological research of the past twenty-five years offers compelling evidence that the etiology of schizophrenia is, to put it succinctly, biological rather than psychogenic" (29). Now, our aim is not to resurrect early psychoanalytical theories about the causes of schizophrenia—theories that resulted in mothers (in those days, caregiving was nearly always provided by mothers) being blamed for their children's poor mental health. However, twenty years after Willick wrote, most experts would not be so definite in their conclusions. As we have already stated, adverse childhood experiences are now seen as one possible cause of schizophrenia (Vallejos et al. 2017; see also Hirt et al. 2019).

Nowadays, medication is the cornerstone of schizophrenia treatment, with antipsychotic medications being the most commonly prescribed drugs (Mayo Clinic 2020). So-called second-generation antipsychotics are less likely to have adverse side effects than the older, first-generation antipsychotics. Much is still unknown about their mode of action, but they are thought to operate by altering brain dopamine levels. Medication obviously operates at the individual level. However, it has been argued that the "chemical imbalance" model that lies behind such pharmaceutical approaches is backed by limited evidence, and that there are no biological tests for the diagnosis of mental disorders (Syme & Hagen 2020). The other major approaches to treating schizophrenia are psychosocial, and while still targeted at individuals or sometimes their families, these take more account of the environment in which people with schizophrenia find themselves. They are therefore less reductionist than approaches that rely solely on medication.

Five main approaches to psychosocial intervention for the treatment of schizophrenia are employed: cognitive therapy (cognitive behavioral and cognitive remediation therapy), psychoeducation, family intervention, social skills training, and assertive community treatment (Chien et al. 2013). It is fair to say that while these don't provide any sort of "magic bullet," they often help (e.g., Stanley & Shwetha 2006; Lim et al. 2017). Overall, it seems that the most efficacious treatments presently available for the treatment of schizophrenia are ones that draw on approaches from several levels, combining medication, individual therapy, and social support in the form of interventions that take account of the relationships that the person with schizophrenia has with their family and others.

Indeed, some recent approaches to tackling schizophrenia overtly refer to "holistic management." In a 2018 article titled "Holistic Management of Schizophrenia Symptoms Using Pharmacological and Non-pharmacological Treatment," Ganguly, Soliman, and Moustafa concluded that:

> Although medications play a role, other factors that lead to a successful holistic management of schizophrenia include addressing the following: financial management, independent community living, independent living skill, relationship, friendship, entertainment, regular exercise for weight gained due to medication administration, co-morbid health issues, and day-care programmes for independent living. (Ganguly et al. 2018, 1)

In the above four case studies—sickle cell disease, deafness, obesity, and schizophrenia—we have examined not only the benefits but also the shortcomings of, on the one hand, reductionist / mechanistic approaches and, on the other, organismic / holistic ones. Before we leave the topic of health, it is worth examining what so-called holistic medicine—an approach that all but eschews mechanism / reductionism—has to offer for human health.

HOLISTIC MEDICINE

Most of us in the West are used, if something goes wrong with our health to the extent that we seek professional expertise, to having a doctor first find out which bit of us has gone wrong and then focusing their attention on that bit. As we get older, or if we end up in hospital, that can often mean dealing with a number of doctors, each with their own specialism, as a number of bits of us fail to work as they should. This is both a reductionist and a mechanist approach. The human body is reduced to its constituent parts and envisaged as a complicated machine with lots of interconnected components. The key to successful treatment is to find the components that aren't working as well as they should be and then deal with them.

Holistic medicine operates within a different understanding of health:

> Holistic medicine is a form of healing that considers the whole person—body, mind, spirit, and emotions—in the quest for optimal health and wellness. According to the holistic medicine philosophy, one can achieve optimal health—the primary goal of holistic medicine practice—by gaining proper balance in life. (WebMD 2020)

How you react to this account of holistic medicine probably says quite a bit about you. Some people react positively, feeling that it encapsulates a dissatisfaction they have felt with conventional medicine. Others reject such a vision. We rarely cite Wikipedia, but it can be useful, as it often, given its public editing facility, provides something of a consensus about issues. Interestingly, at the time of writing, there doesn't seem to be a Wikipedia entry for "holistic medicine"—which is pretty remarkable in itself, given the millions of website hits there are for the term. Instead, here is the start of the Wikipedia entry for "alternative medicine":

> Alternative medicine is any practice that aims to achieve the healing effects of medicine, but which lacks biological plausibility and is untested, untestable or proven ineffective. Complementary medicine (CM), complementary and alternative medicine (CAM), integrated medicine or integrative medicine (IM), and holistic medicine are among many rebrandings of the same phenomenon. Alternative therapies share in common that they reside outside medical science, and rely on pseudoscience. (Wikipedia 2021a)

So, holistic medicine is a rebranding of alternative medicine, lacks biological plausibility, and relies on pseudoscience. For a considerably more positive Wikipedia entry on a contentious subject, try looking up Satanism.

In Western intellectual thought, holistic medicine has grown in prominence since the 1960s as a reaction to the growth of modern, reductionist medicine, but it can trace its roots back to none other than Hippocrates (460–c. 370 BCE), often described as the father of Western medicine, whose students "worked to help their patients to step into character, get direction in life, and use their human talents for the benefit of their surrounding world" (Ventegodt et al. 2007, 1622). In other parts of the world, holistic medicine has a continuous history that goes back thousands of years and has never fallen into abeyance. For example, Ayurvedic medicine is an Indian medical system that dates back more than 3,000 years and places particular emphasis on a "natural" as well as a holistic approach to physical and mental health (National Centre for Complementary and Integrative Health 2021). Treatments rely not only on traditional medicines but also on recommendations for diet, exercise, and lifestyle.

One branch of holistic medicine, anthroposophical medicine, derives from the teachings of Rudolf Steiner (1861–1925), of whom we shall have more to say in Chapter 7 in the context of agriculture. Steiner's ideas about both health and agriculture draw on German idealism and the organicism of Goethe, among other sources. His views on vaccination are complicated—but basically, he was against it:

> And smallpox vaccination? There we find ourselves in a peculiar position. You see, when you vaccinate someone and you are an anthroposophist, bringing him up in the anthroposophical way, it will do no harm. It will harm only those who grow up with mainly materialistic ideas. Then vaccination becomes a kind of ahrimanic power; the individual can no longer rise above a certain materialistic way of feeling . . . Statistics will of course be quoted, and we must ask ourselves if we really must rate statistic so highly exactly in this respect. (Adams 2018)

Today's Steiner Waldorf schools claim that they have nothing to say about vaccination. Here is a statement from Fran Russell, Executive Director of the Steiner Waldorf Schools Fellowship: "School staff are not medically trained, therefore it is outside their remit to advise other staff and families around any medical issues including the decision to vaccinate. . . . The Steiner Waldorf education movement is sometimes linked to suggestions that our schools take an anti-vaccination stance. This is incorrect. Vaccination is a medical, not an education issue and no school is 'anti-vaccination'" (Russell 2021). The reality, though, is that disproportionate numbers of un- and under-vaccinated children attend Waldorf schools (Sobo 2015). There is evidence that COVID-19 infection rates are higher in such schools (Foulkes 2021), although advocates of these schools would argue that their approach leads overall to more balanced and healthier lives.

Traditional Chinese Medicine is another branch of holistic medicine. It has existed for thousands of years and has changed little over the centuries:

> Its basic concept is that a vital force of life, called Qi, surges through the body. Any imbalance to Qi can cause disease and illness. This imbalance is most commonly thought to be caused by an alteration in the opposite and complementary forces that make up the Qi. These are called yin and yang. (Johns Hopkins Medicine 2021)

Treatments to regain balance include acupuncture, the burning of herbal leaves on or near the body, cupping (the use of warmed glass jars to create suction on certain points of the body), massage, herbal remedies, and movement and concentration exercises (such as tai chi; Johns Hopkins Medicine 2021). Conventional medicine has increasingly accepted acupuncture. In the UK, it is used in many NHS GP practices, as well as in the majority of pain clinics and hospices (NHS 2019c). One reason why acupuncture has become increasingly accepted in the West is because a mechanism can be envisaged that "makes sense" in terms of modern science. This mechanism is not couched in the language of the flow of Qi through the body's meridians (as it is in Traditional Chinese Medicine) but rather in terms of the stimulation of sensory nerves that leads to the production of natural substances ("natural" again), such as pain-relieving endorphins (NHS 2019c).

Evaluating holistic medicine is not straightforward. The standard approach in conventional medicine to evaluating a new procedure, whether it's a drug, vaccine, type of surgery, or whatever, is to devise and undertake an randomized controlled trial (RCT). In essence, the new procedure is compared with an existing and trusted procedure (if one exists) or against a placebo (if no existing procedure exists—for instance when devising a vaccine against a new disease). People are then recruited to the trial and randomly allocated to the two possible "arms" of the trial—about half of them getting the new procedure and the other half the existing procedure or placebo. Ideally, neither those receiving the treatments nor the doctors and other staff involved in the trial should know to which arm each person has been allocated—which makes it a "double-blind RCT." This works pretty well when one is testing a new drug, as the new drug can be presented to look the same as the existing drug or placebo. But it obviously doesn't work as well when testing conventional medicine against holistic medicine when it's patently obvious to doctor and patient alike which arm of the trial each patient is in. The reason why this is important is the human mind is powerful; knowing which arm of the trial one is in—and this can be true for the doctor as well as for the patient—can markedly affect the result of the trial. Another issue is that holistic and conventional medicines might have rather different intended outcomes, with holistic medicine likely to give more weight to subjective evaluations—how treated people say they feel.

One of us had a midwife for a mother and an obstetrician and gynecologist for a father. Well, midwifery is about as natural a health profession as you can get. Every human culture has had midwives since before the dawn of history. Fundamentally, what a midwife does is to try to help nature take its course. Interestingly, midwives and obstetricians sometimes have rather different conceptions of what makes for good health, with midwives favoring a more natural, less interventionist approach. A Cochrane review (about as scientifically rigorous as it gets) found that midwifery-led care was associated with a reduction in the use of epidurals, with fewer episiotomies or instrumental births, and a decreased risk of losing the baby before twenty-four weeks' gestation (Sandall et al. 2016). Surgery is very different from midwifery, and largely consists of removing bits of the body that have gone wrong, but it too can be seen as operating within a framework of what is natural. Every surgeon knows that once they have done their bit over a matter of hours ("the operation"), it's up to the patient, with their support mechanisms (family and others), to recover over the weeks and months that follow as their body knits and repairs itself.

THE WAY FORWARD

Are we just saying that everything about human health is very complicated? Not quite. True, when it comes to schizophrenia, there is more that we don't know than do know about causes and what makes for long-lasting, successful treatments. But for the other cases we have examined—sickle cell anemia, D/deafness, and obesity—while things are complicated, there is a great deal about causation and treatment that we do know. Our point is rather that for all four cases, and we suspect for much of human health, if one focuses narrowly on mechanisms and on reducing one's understanding of diseases to the molecular or cellular level, there is much that one misses; equally, if one only looks at the whole person and fails to drill down, there is much that one misses. It is not that medicine should ignore what is going on at the levels of genes, cells, and organs, but rather that it needs also to look at each person as a whole, embedded in a wider environment, and at how their health can be restored to its natural state.

Fortunately, this kind of systems-wide thinking is becoming more common in health services research (Greenhalgh & Papoutsi 2018); indeed, so-called complexity theory is being used in an increasing number of fields (Boulton et al. 2015), including health (Carroll et al. 2021). One feature of a more systemic focus has long been advocated by Denis Noble, who, from a lifetime spent researching the operation of the human heart, has shown the way in which causation operates not only upward (from molecules to organisms) but also downward. Noble (2017) concludes that organisms operate on multiple levels of complexity and must therefore be analyzed from a multiscale perspective. It's rather like a phenomenon that novelists sometimes find—that their stories take on a life of their own. In order to be "true" to the story, the plot and characters develop differently from how the author had expected or even intended, as was the case for Tolstoy who found himself becoming more sympathetic to the eponymous heroine of *Anna Karenina* than he had earlier anticipated. If the fictional products of a novelist's pen can affect their minds, perhaps it's not surprising that biological causation can be top-down as well as bottom-up.

In a link with Chapter 5, we close by emphasizing the importance of our physical environment. It is increasingly being realized how significant this is for our mental as well as our physical health (Johansson et al. 2011; Roe & Aspinall 2011; Palmer 2020; Sturm et al. 2020). Walking, green spaces, thoughtful city planning, and decent architecture can all contribute to reducing stress, increasing positivity, and even addressing the unwanted effects of dementia. If you can't, for whatever reasons, get outside, even watching green spaces on TV or imagining them can help.

Chapter Five

SEX AND GENDER

It is time that we all see gender as a spectrum instead of two sets of opposing ideals. We should stop defining each other by what we are not and start defining ourselves by who we are.

—EMMA WATSON

Identical twins Bruce and Brian Reimer were born in Winnipeg, Canada, on August 22, 1965, sons of Janet and Ron Reimer. At six months, both the twins were diagnosed with phimosis, a common condition among young boys in which the foreskin of the penis cannot retract, inhibiting regular urination. Ninety-nine percent of cases resolve themselves by the age of sixteen, but circumcision is quite often used as a treatment.

Bruce and Brian were both taken, at seven months of age, to be circumcised. Circumcision is usually undertaken as a minor and routine procedure, using a scalpel to remove the foreskin. However, in this case, a cauterizing needle was used, powered by electricity, and the equipment malfunctioned when used on Bruce. (Brian was due to have the procedure after Bruce, but unsurprisingly, he didn't, and his phimosis sorted itself out without any medical intervention.) It is not clear exactly how great the damage to Bruce was, but a surge in current led to his penis either being totally severed or damaged beyond the possibility of function (Gaetano 2021).

This was clearly a personal disaster for Bruce, but what made his case notorious—to the extent that books have been written and TV documentaries made about him—is what happened next. Months went by, and then Janet and Ron Reimer happened to see a psychologist, Dr. John Money, on the TV. Dr. Money, in common with many other professionals at the time, argued that gender was a societal construct and therefore could be changed from an early age. As Janet Reimer put it, "Dr. Money was on there and he was very charismatic, he seemed highly intelligent and very confident of what he was saying" (BBC 2010).

Janet Reimer wrote to Dr. Money at Johns Hopkins Hospital in Baltimore and took Bruce to see him. Dr. Money recommended sexual reassignment surgery and raising Bruce as a girl. Such a practice was not uncommon at the time for children born with genitalia that could not straightforwardly be classified as male or female. Janet and Ron Reimer went along with Dr. Money's recommendation. Bruce's testes and what was left of his penis were removed and a vaginal canal constructed. Following the surgery, Bruce was renamed Brenda and his parents raised him as a girl. During adolescence, he was treated with estrogen (a female sex hormone).

Dr. Money had told Brenda's parents that it was essential that she was never told that she had been born a boy. Brenda and her brother visited Dr. Money annually and, without realizing it, were participating in an ongoing experiment. Brenda was the first known case of sex reassignment of a child born unambiguously belonging to the other sex. Brian served as a control, being genetically identical to Brenda and sharing much the same pre- and post-birth environment (Colapinto 2000).

In publications, Dr. Money stated that the reassignment was a success, claiming that Brenda's girlish behavior was in stark contrast to Brian's boyishness. It has been calculated that these claims of success provided justification for thousands of cases of sex reassignment surgeries. However, Janet Reimer felt differently:

> I could see that Brenda wasn't happy as a girl . . . She was very rebellious. She was very masculine, and I could not persuade her to do anything feminine. Brenda had almost no friends growing up. Everybody ridiculed her, called her cavewoman. She was a very lonely, lonely girl. (BBC 2010)

By the time she reached puberty, Brenda was feeling suicidal. Her parents stopped taking her to Dr. Money, and in 1980, when Brenda was aged fifteen, her father told her what had happened. Within weeks, Brenda had chosen to become David. By the time he was twenty-one, David had had testosterone therapy, breast removal, and surgery to reconstruct a penis. He wasn't able to father children himself, but at the age of twenty-five, he married Jane Fontaine, a single mother of three.

Sadly, though, the story does not have a happy ending. In 1997, David spoke publicly about his experiences for the first time. He described his time with Dr. Money as torturous and abusive. In his early twenties, David had already tried on two occasions to kill himself. His twin brother died from an antidepressant overdose in 2002. On May 2, 2004, David's wife said she wanted a divorce. Two days later, David killed himself.

What is distinctive about the David Reimer story is how it illustrates so acutely the ongoing debate over the relationship between sex and gender and what causes each of us to identify as female, male, or other. So far, it all looks pretty straightforward—David was born male, and the attempt through surgery, upbringing, and hormone treatment to make him female was a failure. Deterministic biology 1, holistic biology 0. But, as we shall see, things aren't quite so straightforward.

SEX DETERMINATION

Most of us at school were probably told that female humans have two X chromosomes and males an X and a Y. It is necessary to specify "humans," as there is an extraordinary range of ways in which nature determines sex in different species. For a start, many species, including most flowering plants and about 5 percent of all animals, are hermaphrodites, so that an individual can produce both female and male gametes—sometimes at the same time ("simultaneous hermaphrodites"), sometimes at different times ("sequential hermaphrodites"). The clownfish that star in *Finding Nemo* are (in nature, not in the film) sequential hermaphrodites. Individuals start off as males, and then the largest in the group becomes a female. If this female dies, its male mate changes sex so that it becomes the dominant female, and the next largest male becomes its mate (University of Exeter 2017). Nemo's dad, Marlin, should have changed into a female (interestingly, Marlin is one of those names like Hilary that is used for both boys and girls—though a feminized version, Marlene, also exists) soon after the start of the film when his "wife" Coral (along with all other offspring except Nemo) got eaten by barracuda—but then there is a long history of nondocumentary films anthropomorphizing animals (Stanton 2021).

Back to humans. All chromosomes under the light microscope tend to look X-shaped, with four "branches," but the Y chromosome is unusual in that two of its branches can appear to be merged under the light microscope—so that the chromosome sometimes looks a bit like a Y. The Y chromosome is unusual in other respects too. It is one of the fastest evolving parts of the human genome, and it has far fewer genes than our other chromosomes, probably only about fifty to sixty, compared to an average of about 1,000 for each of the others.

Unsurprisingly, as only males have Y chromosomes, the genes that are on the Y chromosome tend to play roles in male sex determination and development. One of the most important of these genes is the *SRY* (Sex-determining Region Y) gene. The DNA of this gene, early in development, produces a protein called TDF (testis-determining factor). Classic research by Robin Lovell-Badge and his colleagues showed that when mice *SRY* gene sequences were injected into XX (female) mice early in development, the resulting mice looked male (Koopman et al. 1991).

Until the developing human embryo is about seven weeks old (post conception), there is no visible sexual differentiation. Individuals with XX chromosomes look the same as individuals with XY chromosomes. Then, at about the same time that the embryo ("growing within") starts to be called a fetus ("unborn offspring"), sexual differentiation begins. The TDF protein does several crucial things. For one thing, it begins to act on certain tissues, turning them into what will become the testes. At the same time, it stops the same tissues from developing into ovaries, fallopian tubes, and the upper vagina. And, again at the same time, it turns on genes on other nonsex chromosomes. The net result is that in an XY fetus from about seven weeks of age, levels of the hormone testosterone (which promotes development of the testes) and anti-Müllerian hormone (which stops the ovaries, fallopian tubes, and the upper vagina from developing) are substantially higher than in an XX fetus. In the large majority of cases, such development results in the birth of individuals who are unambiguously considered to be girls or boys.

But sometimes it doesn't. For example, occasionally, during a process in the type of cell division called meiosis that occurs in the production of sperm and eggs, the *SRY* gene ends up on the X chromosome. If this chromosome is in the sperm that gives rise to a fertilized egg and then a baby, the resulting individual ends up with what is called "XX male syndrome." As the term suggests, the individual has two X chromosomes in each of their cells but looks male. Of course, while these X chromosomes look under a light microscope like X chromosomes, the point is that one of them in each cell has the *SRY* gene on it, so these chromosomes differ from the usual X chromosomes. XX male syndrome is rare—occurring in about 1 in 20,000 newborn males. The condition is pretty variable—most affected men are sterile and have small testes, and sometimes their genitalia are ambiguous. In most cases, male secondary sexual characteristics (facial hair, deepening of the voice) don't develop, although this can be successfully addressed by testosterone treatment.

XX male syndrome is only one of a large number of conditions where individuals don't fit unambiguously into the binary classification of "male" and "female" on grounds of appearance (Sandberg et al. 2017). For such a classification of sex to work, one needs a clear-cut alignment of sex chromosomes, sex hormones (such as testosterone and estrogen), and development. An example where these do not align is congenital adrenal hyperplasia. This condition results from a number of different causes but in all cases entails either excessive or deficient production of the hormone androgen by the adrenal glands. Importantly, the gene responsible for the production of this hormone is located on an "ordinary" (autosomal) chromosome (i.e., not the X or Y chromosome). This means that congenital adrenal hyperplasia can affect individuals who are XX *and* it can affect individuals who are XY.

Congenital adrenal hyperplasia results in alteration to either the primary (genitalia) or secondary sexual characteristics. Our secondary sexual characteristics develop at puberty and so include growth of pubic hair and, in females, widening of the hips and growth of breasts and, in males, broadening of the shoulders and enlargement of the larynx. With congenital adrenal hyperplasia, there can be ambiguous genitalia resulting from *in utero* exposure to unusual levels of sex hormones. In some cases, congenital adrenal hyperplasia only comes to light at puberty, which may be unusually early in life, fail to occur, or be irregular.

THE IMPORTANCE OF LANGUAGE

Earlier, when explaining what is meant by "congenital adrenal hyperplasia," we wrote that this results from "either excessive or deficient production of the hormone androgen." It is very difficult (unless one uses exceptionally long-winded language) when writing about variation in human sex to avoid language that can appear judgmental. The words "excessive" and "deficient" imply a "normal" state from which individuals with congenital adrenal hyperplasia depart. Now, "normal" can be used in a statistical sense, as in "someone of normal height," but most of us don't like being described as abnormal. The word is usually not read neutrally, so that someone who is abnormal is often thought to be odd (again, a word that can be read statistically but is often not), even to be morally culpable in some way, however illogical or irrational that is.

This point about words such as "excessive," "deficient," and "normal" applies even more when we are talking about something as intimate as our sex. The term "intersex" is widely used to identify individuals who can't unambiguously be assigned at birth as male or female. While some intersex individuals find the term helpful, and it is the term we use here, others don't. It can be read as reifying the notion of two "correct" sexes, with "intersex" stuck in between in a sort of no-man's (!) land. Terms such as "abnormalities" and "disorders" suffer even more obviously from problems in how they might be heard—one might be abnormally gifted at music and too much order can be obsessional, but on first hearing, which of us wants to be described as abnormal or suffering from a disorder?

Another point about language in this area is illustrated by the phrase "male sex hormones" and "female sex hormones." Although we will use such language (again, to avoid long-windedness or resorting to technical language with which few are familiar), such language itself tends to give the impression not only that people are either female or male but that certain biochemicals are too. The reality is that although, for example, most of us associate testosterone with being male—and this is perfectly valid, as testosterone plays a crucial part in male development, and men typically secrete testosterone at about twenty times the rate that women do—testosterone is not only produced by women but also has a number of important functions in women. Women with low testosterone levels are more likely to be depressed and lethargic, have lower levels of sexual desire and satisfaction, and suffer from muscle weakness. Similarly, estradiol, the main form of "the female" hormone estrogen, is essential in men for spermatogenesis and erection (Schulster et al. 2016).

INTERSEX ISSUES

With these cautions about language in mind, let us go on to consider intersex issues in more detail. We don't need to spend a huge amount of time on what causes some people to be intersex—the short answer is that there can be many causes—but examining causation does indicate something of the complexity of the issue and can help one to realize that things one may have taken for granted in life may not be so straightforward.

XX male syndrome and congenital adrenal hyperplasia are each a cause of someone being intersex. As we saw, XX male syndrome results from a tiny, but crucial, piece of DNA that would normally be on the Y chromosome ending up on the X chromosome. The result is that while the person looks male, their chromosomes look XX. The reason why this matters is that for a person with XX male syndrome, their X chromosomes lack a number of the genes that are normally found on a Y chromosome. Although individuals with XX male syndrome have normal male genitalia and can ejaculate, they are infertile. Indeed, cases of XX male syndrome often come to light only when an individual is referred for fertility evaluation (Butler et al. 1983). Congenital adrenal hyperplasia results in an individual producing either too much or too little of the hormone androgen.

Other causes of being intersex include being XX but one's mother having unusually high levels of testosterone (e.g., because of an ovarian tumor) and being XY but the receptors to the male hormones not working properly (more than 150 different problems with these receptors not working are known, each sometimes called "testicular feminization"). Some individuals have both ovarian and testicular tissue; this used to be called "true hermaphroditism," and the cause is usually unknown, although in some nonhuman animal research, it has been linked to exposure to high levels of certain agricultural pesticides.

Some people restrict the term "intersex" to diagnoses made at birth; others use the terms to include "anomalies" that may appear later in life, often at puberty. Harry Klinefelter was a doctor who studied medicine at Johns Hopkins School of Medicine. After graduation in 1937, he took a job at Harvard Medical School under a physician famous for his work in endocrinology (Callaghan 2009). One of Klinefelter's first patients was a man with enlarged breasts and small testes. In 1942, with colleagues, Klinefelter published an article in which he described nine such men (Klinefelter et al. 1942). Each had enlarged breasts, small testes (despite a normal-sized penis), and unusually long arms.

Klinefelter presumed that the cause was a hormone malfunction, but eventually it was shown that men with what is now called "Klinefelter syndrome" were XXY—they had forty-seven chromosomes rather than the usual forty-six. Individuals conceived with forty-seven chromosomes usually die before they are born or exhibit major differences from other people, as is the case for most people with Down syndrome, where individuals have three copies of one of the smallest of our chromosomes, chromosome 21. People with Klinefelter syndrome, though, are relatively unaffected, probably because of a phenomenon that occurs in women called "X chromosome inactivation" (Kinjo et al. 2020). What happens is that early in the development of a female, one of the two X chromosomes in each cell is prevented from making any proteins. Females therefore functionally only have forty-five chromosomes in each of their cells. Normal males have forty-six (although, as mentioned above, there are very few genes on the Y chromosome). In males with Klinefelter syndrome, each cell behaves pretty much as if it is XY rather than XXY.

There are many other types of intersexuality, and it is difficult to determine its overall extent. If one includes conditions such as Klinefelter syndrome, Turner syndrome (women with forty-five chromosomes, as they have only one X chromosome), and late-onset adrenal hyperplasia, a figure of 1.7 percent has been calculated (Fausto-Sterling 2000). Adopting much narrower criteria, a figure of 0.02 percent has been calculated (Sax 2002). A more recent review by Intersex Human Rights Australia noted many sources had lower bound limits of 1 in 1,500 to 1 in 2,000 (0.05 percent), but also noted that if we use a definition of intersexuality that focuses on genitalia that make classification at birth as a boy or girl difficult, a figure of about 0.6 percent is more reasonable (Intersex Human Rights Australia 2013).

Traditionally, parents could only be told whether their child was a boy or a girl once it had been born. Increasingly, unborn fetuses can have their sex identified. Often, this is just to satisfy parental interest, but sometimes it can be used for medical or more nefarious reasons—there are many countries where female fetuses are more likely to be aborted for "cultural" reasons. Ultrasound is commonly used, as this allows visual identification of the penis, although various genetic tests are also available. The practice has led in some cultures to increasing numbers of gender-reveal parties where the parents-to-be tell family and friends whether they are expecting a girl or a boy.

Media reporting of these gender-reveal parties usually occurs when there is a disaster. For example, in 2020, the use of a pyrotechnic device (a smoke bomb) at one such party in El Dorado, California, with temperatures well above normal, led to a fire that burnt for twenty-three days, consumed more than 9,000 hectares, destroyed twenty buildings (five of them homes), and directly resulted in the death of a firefighter, Charlie Morton. The couple concerned were subsequently charged with one felony count of involuntary manslaughter, three felony counts of recklessly causing a fire with great bodily injury, four felony counts of recklessly causing a fire to inhabited structures, and twenty-two misdemeanor counts of recklessly causing fire to property of another (Canon 2021).

Such parties have also been criticized for embodying stereotypical or essentialist views of gender. Indeed, the woman, Jenna Karvunidis, who seems to have held the first such party (in 2008), celebrating the imminent arrival of her daughter with a pink cake, now sees things rather differently:

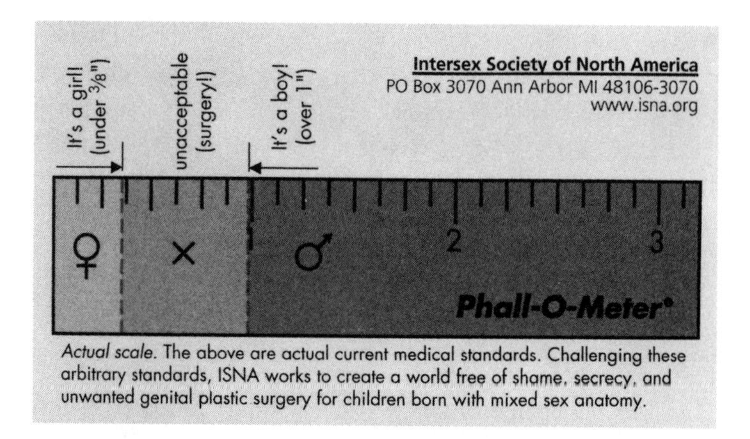

FIGURE 5.1 How to tell if a newborn is a boy or a girl. *Credit:* Wellcome Collection / CC BY 4.0.

> Who cares what gender the baby is? I did at the time because we didn't live in 2019 and didn't know what we know now—that assigning focus on gender at birth leaves out so much of their potential and talents that have nothing to do with what's between their legs . . . Gender reveals are really offensive to nonbinary and transgender people. When we emphasize gender as the first thing to celebrate about babies, [transgender and nonbinary people] are further marginalized. (Harmsen 2019)

But how precisely does a midwife or doctor tell if a newborn is a boy or a girl? A quick inspection of the genital region usually suffices, but, a common trope in this book, things are not always so straightforward. The Phall-O-Meter (Figure 5.1) is a semi-satirical device devised by Kiira Triea, one of the founders of the intersex movement, based on a book by Suzanne Kessler (1998) in which she summarized existing views on what were considered medically acceptable infant penis and clitoris sizes. In reality, of course, medical practitioners and others used more subjective approaches. For a more rigorous approach, try looking up the Prader scale or Quigley scale.

What matters, though, is what happens to newborns who cannot straightforwardly be classified as female or male. As the labeling on the Phall-O-Meter indicates, a common response has been surgery (Dickens 2018). Indeed, Blackless et al. (2000) estimated that the incidence of "corrective" genital surgery in the United States was about 1 to 2 per 1,000 births (0.1 percent to 0.2 percent). There are two intertwined issues here: an ethical issue to do with consent, and a conceptual issue to do with what is deemed to be desirable. A worrying feature, from today's perspective, is that much of the early surgery was undertaken without any parental consent. Doctors, and perhaps the general culture at the time in the early second half of the twentieth century, simply presumed that what a doctor thought was appropriate was good enough.

In 2001, a pediatric endocrinologist, Jorge Daaboul, and an ethicist, Joel Frader, argued for what they described as "a middle way":

> Traditional practice involves paternalistic decision making by medical practitioners, including the use of deception and/or incomplete communication of facts about the infant's condition and early surgical intervention to make a "definitive" sex and gender assignment. However, modern scientific evidence about sex-role determination refutes earlier theories supporting the appropriateness and need for early decisions. Some intersex individuals have begun to speak out against their treatment, denouncing the secretive approaches and cosmetic surgery without the specific consent of the (mature) affected individuals. They argue for complete disclosure of information regarding the condition and deferral of all surgery until at least adolescence. The traditionalist practices no longer conform to modern legal or ethical standards of care. The position of some intersex activists ignores the potential for psychosocial harm to intersex children and our society's general and strong deference to parental discretion in decisions for and about their children. We argue for a middle way, involving shared decision making with parents of children with intersex and the honoring of parental preferences for or against surgery. (Daaboul & Frader 2001, 1575)

Here, Daaboul and Frader reject a growing demand from the intersex movement that surgery should not be undertaken on newborns simply because they don't conform to general presumptions about what a baby should look like. There are similarities here to foot-binding among young girls and what is now generally called female genital mutilation (also known as female genital cutting and female circumcision). We don't need to go into great detail here. Both practices were (in the case of female genital mutilation still are in some places) deemed culturally necessary and led to many hundreds of millions, possibly billions, of women leading lives that in terms of physical comfort and pleasure were (are) very substantially diminished. Foot-binding died out in the twentieth century, and female genital mutilation is becoming less common (United Nations Children's Fund 2016).

The surgery required for intersex individuals is not straightforward (for details, see Callaghan 2009). In the case of what are called "feminizing surgeries," what usually happens is clitoral reduction and vaginoplasty (construction or enlargement of a vagina). Vaginoplasty is not an easy surgical procedure, but "masculinizing surgeries" make it look relatively straightforward. The problem is building a conventional penis when you don't have that much to start with. It is difficult to draw firm conclusions about how successful these operations are. In many cases, data on satisfaction come predominantly either from follow-ups by doctors with former patients or from members of intersex support groups. Follow-ups by doctors with former patients often suggest that the surgery was welcomed; support group data point in the opposite direction. This may tell us as much about who goes to support groups and how easy it is for people to tell their doctors that the surgery was a failure as anything else.

A German working party on medical management of differences of sex development (DSD) / intersex in early childhood drew on three ethical principles:

> (1) to foster the well-being of the child and the future adult, (2) to uphold the rights of children and adolescents to participate in and / or self-determine decisions that affect them now or later, and (3) to respect the family and parent-child relationships. (Wiesemann et al. 2010, 671)

This is not dissimilar to the earlier conclusions of Daaboul and Frader (2001) but gives more weight to the views of children and adolescents. The direction of travel is clear, and there is a growing swell of voices from those who reject surgical intervention (e.g., Open Society Foundations 2019). In 2015, Malta banned intersex surgery for children too young to consent. It seems clear that the changes over the last half century have resulted partly from a growing belief that doctors do not necessarily know best and that, by and large, patient consent is needed, barring medical emergencies, and partly from a loosening up of attitudes so that more people are now prepared to accept that the world is not neatly divided into "males" and "females" but that these states lie at opposite ends of a spectrum. This leads on to a consideration of gender.

GENDER

The same Dr. John Money whom we met at the start of this chapter was responsible in the 1950s for the terms "gender identity" and "gender role" (Ehrhardt 2007). Money argued that a distinction could be made between a person's biological sex and their gender—a notion that was subsequently enthusiastically adopted by the growing feminist movement in the West. Money underestimated the importance of biology, understood narrowly, as David Reimer's story indicates, but the distinction he drew between biological sex and gender has been very influential. Our position is different from his: the relationship between sex and gender is neither one of determinism nor one of independence. Our gender develops from our sex and rests upon it, but, a common theme in this book, what we see at one level in biology—gender, in this case—while affected by the level beneath—sex, in this case—is not determined by it. Gender cannot be reduced to sex.

Simone de Beauvoir studied mathematics and philosophy at the Institut Catholique de Paris and the Sorbonne. At the age of twenty-one, she was the youngest person ever to pass the prestigious *agrégation* in philosophy, coming second in the nationwide examination, by a narrow margin, to Jean-Paul Sartre (Mussett 2021). She spent the next sixty years, until he died in 1980, with him (and a number of others, male and female), and it is somewhat ironic, for as independent a woman as she was, that she has often been seen in his light. Sartre was a wonderful novelist (cf. *Nausea*) and deservedly won the Nobel Prize for Literature, but while a leading existentialist, anyone who has read all of his *Being and Nothingness* is likely to find it considerably less engaging and original than de Beauvoir's *The Second Sex*.

The Second Sex was published in 1949. It is often described as the starting point of second-wave feminism, first-wave feminism having previously concentrated on legal matters, such as the right to own property and to vote. Second-wave feminism is all about broadening the debate to issues to do with the family, sexual expression, reproductive rights, and the workplace. De Beauvoir's *The Second Sex* looks at all of these—but is rooted in a philosophical argument that man is considered to occupy the default position, whereas woman is always "The Other." De Beauvoir looked at the issues from every angle; indeed, she even spent time in her book considering the relationships between eggs and sperm in various species, including humans. Acknowledging the relevance of biology, she nevertheless concludes "One is not born, but rather becomes, a woman"—a sufficiently radical conclusion that it helped get *The Second Sex* on the Vatican's list of prohibited books. (One wonders if de Beauvoir took that as a backhanded compliment. As a teenager, she had seriously considered, like any good convent-educated girl of the time, becoming a nun. She soon abandoned the idea and became a lifelong atheist.)

At the same time as de Beauvoir was writing about eggs and sperm, Ruth Herschberger (1948) was arguing that that the female reproductive organs are viewed as somehow being less autonomous than those of the male. Emily Martin has shown that while menstruation is typically seen in biology textbooks as a failure (you should have got pregnant), sperm maturation is viewed as a wonderful achievement in which countless millions of sperm are manufactured each day (Martin 1991). Sperm are viewed as active and streamlined, whereas the egg is large and passive and just drifts along or sits there waiting. A rare exception (the exception that proves the rule) is provided by Woody Allen's 1972 *Everything You Always Wanted to Know About Sex* (*But Were Afraid to Ask)*. Here, Allen plays the part of an anxious sperm, worried by approaching orgasm when he will be launched into darkness, not sure if he will be trapped in a contraceptive device or where he will end up if the man masturbates.

The way the human egg is portrayed in biology textbooks has been likened to that of the fairy tale *Sleeping Beauty,* in which a dormant, virginal bride awaits a male's magic kiss. This is despite the fact that academic biologists see both egg and sperm as active partners. Just as sperm seek out the egg, so the vagina discriminates between sperm, and the egg seeks out sperm to catch. Nevertheless, as Martin points out, even when acknowledged, such biological equality is still generally described in a language that gives precedence to the sperm. When the egg is presented in an active role, the image is one of a dangerous aggressor "rather like a spider laying in wait in her web" (Martin 1991, 498).

The point of all this is that whether you see the egg as dormant and sperm as active or the egg as wise and sperm as indiscriminate is far more to do with human culture than biology understood narrowly. (We keep on having to add "understood narrowly" because to us, biology is not just the basics; it's about the totality of organisms, including humans. One of us used to teach the subject "social biology" to sixteen- to nineteen-year-olds. It was a great subject with everything from human anatomy and physiology through to human behavior, including the history of tool use and an introduction to both psychology and sociology.)

One point where human sex and gender obviously intersect is motherhood. One day, we may perhaps have the option of babies being grown in "artificial wombs" (exowombs—cf. The Central London Hatchery in Aldous Huxley's *Brave New World*)—and fetal lambs that are developmentally equivalent to extreme premature human infants can be physiologically supported in an extra-uterine device for up to four weeks (Partridge et al. 2017). But at present, as every mother knows, it's more or less nine months of carrying the developing baby, excepting surrogacy, which as yet is (virtually) always undertaken for medical (infertility) rather than social (convenience) reasons. After the end of pregnancy, another feature of parenthood to do more with biological sex than gender comes into play—breastfeeding. Of course, breastfeeding in many cultures typically only lasts a matter of months, although in some, it lasts much longer: in nonindustrialized societies, the average duration is twenty-nine months (Veile & Miller 2019). And fathers can undertake bottle-feeding (including bottle-feeding with breast milk), but it's hardly surprising that for most couples, it's the mother who takes the lead on feeding. Given the fact that most societies that have parental leave give more time to mothers than to fathers, it is, again, hardly surprising that it's mothers who end up doing most of what is needed to raise a child—and not just until the end of breastfeeding.

Cultural forces are powerful—just as those to do with biology (again, understood narrowly) are. In countries where parental leave is transferable between the mother and father, it's rare for the father to take as much leave as the mother. In 2020, Finland (always a bastion of progressive gender policies) introduced a system whereby all mothers and fathers will each get nearly seven months' paid leave, half of which will be nontransferable (Topping 2020). It will be interesting to see the result. As many women know, legislative attempts to reduce gender inequalities often fail to produce their intended changes in human behavior.

Wombs and breasts are one thing, but brains are another. There has long been debate as to whether there are important differences between women's and men's brains. According to one report (CNN 1999), the top-selling nonfiction book of the 1990s (scriptures are typically excluded from such lists) was John Gray's (1992) *Men are from Mars, Women are from Venus.* To date, it has sold some fifteen million copies. Its core argument is that relationship difficulties between men and women result from fundamental psychological differences between the sexes. It is easy to be patronizing about the book, and thirty years after its publication, some of it is pretty cringeworthy. But it raises a crucial question: do men and women have distinct personalities, and if they do, why is that?

Let's cut to the chase—the current academic consensus, with which we agree, is that (i) there are no clear-cut mental differences between men and women of any significance, and (ii) what differences there are have more to do with culture than biology. The evidence for the first of these conclusions—that there are no clear-cut mental differences between men and women of any significance—comes from innumerable empirical studies in which men and women fill out various surveys (e.g., Carothers & Reis 2012). It is possible to claim that such surveys miss what is important in human psychology or that people don't respond truthfully, but the fact remains that the best data we have show little or no sex differences, not just with regards to such personality traits as introversion versus extroversion and how conscientious one is, but also with regards to questions about empathy, sexual attitudes, and desire. This is in marked contrast with empirical studies on things such as physical strength where, although there is very considerable overlap, it is meaningful and statistically valid (despite exceptions) to make distinctions between the average male and the average female.

Just as importantly, what sex differences there are in mentality are to do more with culture than biology. We can start by looking at what physical differences there are between the brains of men and of women; this is a pretty crude way of looking at sex differences, but it's worth starting with it for two reasons. First, as we all know, there are important and relevant differences between male and female anatomy with regards to both primary and secondary sexual characteristics, as discussed above. It doesn't seem possible to rule out *a priori* the possibility of sex-specific differences in respect of brain anatomy. Second, if there are such differences, it might be that any attempts to reduce the influence of sex on mental activities (principally, thoughts and feelings) would be harder, just as we argued above that attempts to reduce the influence of sex on the pre- and postnatal care of babies is difficult because of important and relevant differences between male and female anatomy (wombs and breasts).

In the event, and despite frequent arguments in popular culture and (more rarely nowadays) occasionally in the academic literature that sex differences in the brain do exist and are important, the strong scientific consensus is that this is not the case. Early studies that argued the opposite had very small sample sizes or, even by the standards of the time, were often poorly carried out—sometimes being the product of not much more than wishful thinking (Gould 1981). Some studies even failed to take account of the fact that women often have smaller brains than men simply because women's bodies are typically smaller.

We are still a long way from measurements of the brain being able to tell us much of interest about the mind, but recent studies, for example from magnetic resonance imaging scans, conclude that sex differences are small or very small (Joel et al. 2015; Eliot 2019; Rippon 2019). Furthermore, it needs to be emphasized that such studies are correlational studies, obviously undertaken on individuals who have lived their entire lives in societies that generally presume to a greater or lesser (Finland again) extent that men and women do and should think and behave differently. Such sex-specific differences in brain anatomy that do exist might therefore be the product rather than the cause of these cultural differences, just as licensed London taxi drivers have posterior hippocampi that are significantly larger than controls (Maguire et al. 2000). This is presumably because such taxi drivers are highly dependent on their navigation skills, having to pass a demanding test ("The Knowledge") that requires exceptionally detailed memory of London streets without recourse to satnav software. It is in theory possible that individuals with large posterior hippocampi are more likely to become licensed London taxi drivers, but the finding that the longer they have been drivers, the larger their posterior hippocampi are strongly suggests that this is a case of an environmental difference (one's job) resulting in an anatomical effect (cf. the proverbial blacksmith's arm muscles).

What seems to be going on is that the differences between males and females that do result from biology—above all, being pregnant and breastfeeding—lead to cultural assumptions about which sex should do most of the looking after children. This is magnified by the fact that in many societies, married men worry that they might be cuckolded, and parents worry that their daughters might become pregnant before they marry, and so women of childbearing years often enjoy fewer freedoms than their male counterparts. Daughters and sons are, for these and other reasons, therefore likely to be brought up differently; as Lise Eliot (2009) puts it in the subheading to her book *Pink Brain, Blue Brain: How Small Differences Grow into Troublesome Gaps*. Interestingly, until children are about twelve to eighteen months old, there don't seem to be consistent sex-related preferences for toys. But this then changes. As Eliot puts it:

> In dozens of studies, parents have been found to respond more positively when their young child picks up a sex-appropriate plaything, such as when a boy uses a hammer or a girl pushes a toy shopping cart. And they are likelier to bristle when their child plays with the "wrong" type of toy, like a boy cuddling a baby doll or a girl brandishing a toy sword. Fathers react more strongly than mothers, especially when they see their sons engaging in typical female play. By four years old, boys themselves voice the awareness that their fathers would think it "bad" if they played with toys for girls, even something as seemingly neutral as toy dishes, which the boys in this study found quite attractive. (Eliot 2009, 111)

Gender illustrates the complex interplay between features of our biological sex, always remembering that not everyone is either a boy or a girl, and how we behave and see ourselves. Transgender issues add another layer of unwanted complication for those seeking a simple story based on an essentialist understanding of human nature.

SEXISM

Sexism is prejudice or discrimination that results from a person's sex or gender. It mostly negatively affects women or girls, whether transgender or not, although boys and men can be affected, for example in the not uncommon assumption that women are more suited than men to be elementary school teachers or midwives. Indeed, many countries do not allow men to train as midwives, and in those that do, it is rare. In the UK, men have been allowed to be midwives since 1983 (the Sex Discrimination Act was passed in 1975), but fewer than 1 percent of midwives are men in the UK (Nursing and Midwifery Council 2016) and in the United States (Downing 2016), as well as in many other countries. However, in some Southeast Asian cultures, the majority of midwives are men (Gianno 2004).

There may be some readers at this point who are thinking "Typical! A book written by two men and in the section headed 'Sexism,' we are being treated to data on the underrepresentation of men in midwifery." The reality, of course, is that it is overwhelmingly women who are disadvantaged by sexism, whether with regards to education, careers, harassment, rape, or a pervasive culture that all too often presumes that girls and women are simply less capable or worthy of consideration and respect than are boys and men.

Sexism is not only the result of the thoughts and actions of individuals, but it can also be institutional—embedded in organizations—or structural (systemic)—embedded in societies. Structural sexism is particularly evident in language. At one time, those professionals who fought fires were inevitably referred to as "firemen." This is not a good way to encourage women to join the profession. In 2017, the London Fire Commissioner, Dany Cotton, asked members of the public to show their support by stating that they would not write "fireman" on their social media accounts, using the hashtag #FirefightingSexism. As he put it:

> The first woman firefighter joined London Fire Brigade in 1982 and it's ridiculous that 35 years later people are still surprised to see women firefighters or calling them firemen. London is a complex and challenging city and it takes a diverse selection of skills, strengths and specialisms to

> protect it—qualities that both men and women possess. I want to shake off outdated language which we know is stopping young girls and women from considering this rewarding and professional career. (London Fire Brigade 2017)

Ameliorating, let alone overcoming, sexism is not easy. It takes time and requires efforts at all levels, from individuals through organizations to national governments and beyond. Nevertheless, despite instances where things go backward, without wanting in any way to be complacent, there have been major achievements over the last 200 years, though much still remains to be done. A very common trope in feminist science fiction writing, from Margaret Cavendish's 1666 *The Blazing World* through Ursula K. Le Guin's 1969 *The Left Hand of Darkness* and Joanna Russ's 1970 *The Female Man* to Margaret Atwood's 1985 *The Handmaid's Tale* and such contemporary works as Naomi Alderman's 2016 *The Power,* is to imagine how things might be. Some of this writing is utopian, some dystopian.

TRANSGENDER ISSUES

Alexander Walker is a teacher from Belmont, United States:

> Identity is an aspect of one's character that never stops evolving. For most of my life, I felt like I was sitting in no man's land, waiting for someone to give me a push so I could finally feel whole.
>
> When I [was] first exposed to the concept of gender identity and transgender, I was terrified. It was like a tidal wave washed over me saying, "Pay attention to this!" For years, I worked really hard to ignore it. I came to a point where I could not ignore that feeling any longer and I got the push I needed. I began looking for stories about others like me, searching for information and reassurance that I would be okay.
>
> As a young elementary teacher, the lack of visibility and horror stories I found of trans teachers being fired pushed me further and further from my desire [of] living authentically. I sat at a crossroads; should I risk it all—my career, my family, my relationship, my friends—to be myself or continue hiding and see what happened? With support, encouragement, and a lot of courage, I found my own way to and through transition, finding other

> trans people with stories similar to mine along the way. Today, no man's land is a distant memory, and I sit here with more confidence than any other time in my life. (*The New York Times* 2020)

There are as many transgender stories are there are transgender people. Until recently, the condition was little known in most societies. Transgender individuals, while sometimes surprised and encouraged by the support they receive from family, friends, and work colleagues, too often face stigmatization, discrimination, and violence (American Psychological Association 2009). Someone is transgender if they identify with a gender other than that associated with the sex that they were assigned at birth. Iconically, this means someone being born male but from an early age feeling (realizing) that they are a girl or, like Alexander Walker, being born female but from an early age feeling (realizing) that they are a boy. Trans people are said, in medical language, to manifest gender dysphoria—strong persistent feelings of discontent with one's assigned gender and identification with another (or no) gender that result in significant distress and impairment (American Psychiatric Association 2013).

Some trans people reject the term "gender dysphoria," and the term "gender diversity" is sometimes used more generally to reject a binary classification of people into female versus male, and instead describe the wide range of gender identifications outside conventional gender categories. At the present time, arguments about transgender and other gender diversity issues can be passionate to the point of violence (Davy 2021). The arguments become especially heated when it is children who want to transition (Shrier 2020; Stock 2021). Try looking up the debate around "trans-exclusionary radical feminism" (a term generally rejected by those so labeled). As one recent article put it:

> Gender diversity is a topic that generates strong reactions and often polarised views. Claims to certainty about origins and management are common, despite the often very limited empirical basis of such claims. Arguments over the cause, meaning, stability and significance of these gender-diverse feelings are engaged with in deeply anxious ways in our changing society, especially given that we are nowadays equipped with the technologies to make significant bodily interventions. (Wren et al. 2019, 351)

Trans issues are on the rise (Halberstam 2018; Suissa & Sullivan 2021), but it is important to recognize that many cultures have long rejected a binary classification of people, in which everyone is either female or male (Dea 2016). Examples of the many millions of people who do not fit into "the standard gender model" include the Hijra of the Indian subcontinent (where they are officially recognized as third gender), "two-spirit" people in some native American tribes (who sometimes reject the label), the *māhū* (meaning "noble" or "in-the-middle") on a number of Pacific Islands, including Hawaii and Tahiti, and the *mukhannathun* in pre-Islamic and early Islamic times.

From our point of view, gender diversity illustrates a number of themes. First, the existence of gender diversity does not fit with a narrow, essentialist reading of sex and gender in which there is a one-to-one correspondence between a person's sex (understood as a binary) and their gender (also understood as a binary). Second, it illustrates how, as humans, we are neither independent of our biological heritage nor reduced to it. Third, it highlights how part of being human is to be able to exercise autonomy and live authentically. These are themes to which we return in Chapter 6 on race.

Chapter Six

RACE

> Race and racism is a reality that so many of us grow up learning to just deal with. But if we ever hope to move past it, it can't just be on people of color to deal with it. It's up to all of us—Black, white, everyone—no matter how well-meaning we think we might be, to do the honest, uncomfortable work of rooting it out.
>
> —MICHELLE OBAMA

EIGHT MINUTES FORTY-SIX SECONDS

On May 25, 2020, George Floyd, a forty-six-year-old Black American man, died in Minneapolis while being arrested on suspicion of having used a fake $20 bill to buy cigarettes (BBC 2020; Wikipedia 2021b). During the arrest, Derek Chauvin, a White police officer, knelt on Floyd's neck for nine minutes and twenty-nine seconds while Floyd was handcuffed and lying face down. The duration was initially reported as eight minutes and forty-six seconds because a bystander's video began with Chauvin's knee already on Floyd's neck, and it was only corrected subsequently with the release of police body camera footage. The phrase "eight minutes forty-six seconds" has since become a symbol of racial injustice or police brutality, and there have been many "die-ins" where people lie still for this period of time in protest. Video footage showed Floyd repeatedly saying

"I can't breathe," begging "Please, please, please," gasping "You're going to kill me, man," and then saying "Can't believe this, man. Mom, love you. Love you. Tell my kids I love them. I'm dead." During the final few minutes, Floyd was motionless, and no pulse could be found, while Chauvin ignored onlookers' pleas to remove his knee, which he did not do until medics who had arrived told him to.

Derek Chauvin was convicted of two counts of murder and one of manslaughter and is currently serving a twenty-two-and-a-half-year prison sentence. Thomas Lane, who helped restrain Floyd, pleaded guilty to a state charge of aiding and abetting second-degree manslaughter and has been sent to prison for three years. At the time of writing, the cases against officers J. Alexander Kueng, who also helped restrain Floyd, and Tou Thao, who prevented bystanders from intervening, have not yet gone to court. What, of course, propelled Floyd's death from a local incident into a worldwide series of protests that gave powerful impetus to the Black Lives Matter movement was the fact that Floyd was Black and Chauvin is White, and that videos of the period leading up to Floyd's death and subsequently were captured on a number of mobile phones and rapidly shared through social media. This chapter is about words such as "Black" and "White." Is their use valid? And how are we to understand such words if they are valid? Our argument is that answering these questions and addressing the whole issue about the validity of "race" as applied to humans is helped by keeping in mind the two very different approaches of mechanism and organicism that we are using as our two poles throughout this book.

RACE AS A BIOLOGICAL TERM

Race as a concept is controversial not just in politics and broader human culture but also within biology. In an attempt to start from a place where there is at least some degree of consensus, we begin by considering how the term "race" is used by biologists when discussing nonhuman species. Imagine a species; it doesn't much matter what sort of species it is, but to help focus, let's imagine it is a common rodent found over quite a large geographical area—something like a squirrel.

Our squirrel species consists of a number of individuals all alive at the same time. If we were paleontologists, studying fossils, we would also be interested in squirrels that have gone to meet their maker, but we obviously wouldn't be able to find out as much about them from their remains as we can from studying living squirrels. Individual squirrels differ from one another in all sorts of ways, and biologists have to decide where one species of squirrel stops and another starts. Books have been written, by both biologists and philosophers, on the issue of how to define species, but there are three principal ways. The oldest and most straightforward way is to concentrate on morphology—the size, structure, and shape of squirrels in this case. For example, grey squirrels, with the scientific name *Sciurus carolinensis,* can easily be distinguished from red squirrels, *Sciurus vulgaris.* Grey squirrels (there is a clue in their name) generally have grey fur (though they can be brown or black) and, when adult, tend to have a mass of between 400 and 600 g. Red squirrels generally have red fur (though they can be dark to be point of almost being black) and, when adult, typically have a mass of between 250 and 340 g.

The second way to see if two organisms belong to the same species or not is to see if they can mate and produce viable young that themselves can go on and breed successfully. Now, grey squirrels are native to the United States and Canada, and red squirrels to Eurasia. So, before grey squirrels were introduced to the UK in the nineteenth century—the 11th Duke of Bedford, Herbrand Russell, gets much of the blame (Signorile et al. 2016)—and to mainland Europe in 1948, they didn't meet in the wild. In fact, even when they do meet up, they don't mate, let alone produce viable young. So, grey and red squirrels meet both the morphological and the interbreeding criteria used to distinguish species.

The third approach has only been possible in recent decades, and that is to use some sort of molecular means to see how similar putative species are. Technically, this approach can be seen as a subset of the approach that relies on morphology. Nowadays, analyses of DNA are almost invariably used, but in the early days, other approaches were employed, such as analyzing proteins.

Most biologists, if push came to shove, would probably say that the second approach to deciding what a species is—using the criterion of interbreeding—is the one that is closest to what is wanted. However, this criterion has a number of disadvantages. For a start, it can't be applied to fossil species (and most species that have existed are extinct by now) or to species that are asexual. Furthermore, as a criterion, it's pretty impracticable, even for common species—how long do you have to sit around with binoculars waiting to see if any red and grey squirrel pairs head off into the bushes, where it may anyway be difficult to see what precisely is going on, and to determine if healthy offspring subsequently result? In practice, the majority of species identifications are undertaken using the first of the three criteria: morphological similarity.

The individuals that make up a species vary in all sorts of ways, and it is this collection of individuals that characterizes a species. Biologists sometimes, less often nowadays than was once the case, refer to the "wild type" of a species. The wild type is the "normal" or "typical" version. But as we saw in Chapter 4 in relation to sex, the word "normal" is a slippery one that all too easy morphs from the descriptive ("you are much less tall than the average") into the evaluative ("you are abnormally short") and then the judgmental ("you are too short") with language that may offend ("you are a dwarf").

Furthermore, the idea of the normal or wild-type version of a species can be profoundly misleading and, in many ways, runs counter to the whole idea of evolution. There is no Platonic grey squirrel—the transcendent (aspatial, atemporal) Form of a squirrel—hidden from mortal sight that defines what a squirrel is or should be. What you see is what you get. Evolution does not strive to produce a particular sort of squirrel; there is no squirrel blueprint or representation against which each squirrel can be compared. Rather, the blind forces of evolution—we are one with Richard Dawkins (1986) here—constantly do their best to adapt the endlessly shifting collection of squirrel individuals to the specific circumstances in which squirrels find themselves.

And here is where we start, via squirrel subspecies, to get to squirrel races. The grey squirrel can be divided into five subspecies. Whereas all squirrels in the UK belong to the one subspecies, *Sciurus carolinensis carolinensis,* which is white on its underside, all five subspecies are found in different parts of Canada and the United States. For example, *S. c. carolinensis* is found across much of the southern United States except for the wet forests of the Mississippi River delta where *S. c. fuliginosa,* a darker form often with cinnamon on its underside, occurs (Koprowski et al. 2016).

In general (not just with respect to grey squirrels), different subspecies can interbreed successfully but can be distinguished from one another by slight differences in morphology. Things tend to get a bit subjective at this point: when is a difference in morphology sufficient to decide one has got a subspecies? The honest answer, as in quite a bit of science, is "when you can persuade your peers"—peer review is central to science. This degree of subjectivity (which often upsets those who are not professional scientists and assume that science is only about objective accounts of the world) is also found in the taxonomic hierarchy above the species level: it isn't really possible objectively to decide where one genus ends and the next one begins, and the same point holds with families, orders, classes, phyla, and even kingdoms.

When we go to levels below subspecies, things get even less clear. Traditionally, races were recognized as the level beneath subspecies, but the term is an informal one in the sense that it is not defined by professional taxonomists and is even more subjective than are subspecies. Pretty much all that was needed to claim that a species or subspecies could be divided into races were *some* sort of criteria that allowed one race to be distinguished from another. In some branches of biology, "strains" are recognized (again, informally) at a level beneath races. Viruses, which are a bit of a law to themselves, being parasites and not really recognized as living, are classified, much like other species, into species, genera, families, and orders. However, "It is *de facto* accepted by the virologists that there is no homogeneity in the demarcation criteria, nomenclature and classification below the species level, and each specialty group is establishing an appropriate system for its respective family" (Fauquet et al. 2008, 784). Virologists do refer to strains. So, severe acute respiratory syndrome–related coronavirus is a species with a number of strains, one of which is severe acute respiratory syndrome coronavirus 2 (SARS-CoV-2)—the virus that causes COVID-19. As most of us know from COVID-19, virologists are likely to use the word "variants" in discussing important variation at a level beneath that of strains.

Subspecies, races, and varieties can interbreed successfully within their species. So, what is needed to define them within a species is some sort of physical distinction or set of distinctions whether visible or only in evidence at the molecular level (the criterion of interbreeding only being of use to distinguish species rather than higher- or lower-level taxa). Taxonomists—who get paid to do this sort of stuff—are traditionally divided into "splitters" and "lumpers." Splitters are more likely to identify different categories, lumpers to eschew them. To biologists, what is more important than whether races or other subspecific variants are named is the fundamental idea that species change over time. Sometimes, they change very little, but at other times, particularly when there are changes in the environment, they change more rapidly. As Darwin realized in his *Origin* (1859), over time, a species may divide into two or more new species. Races and other subspecific variants are an intermediate stage in this process of the evolution of new species.

That having been said, the term "race," while still used in biology, is nowadays used less than it used to be. This is partly because of the mess that biologists and others got into when using the term in relation to humans.

HUMAN "RACES"

In much of biology, while there are academic disagreements about the specifics, the fundamental idea that in many species a single species can be divided up into distinct subcategories isn't that controversial. Indeed, it can be very useful. For example, it is common to find that different races of fungal pathogens have very different effects on the crops that they attack. Such information can be helpful when dealing with outbreaks of plant diseases.

However, when we come to humans, we are operating in a very different context—to the extent that when the term "race" is applied to humans, it is not uncommon to see it placed in inverted commas (scare quotes), as we have done in the heading to this section, signaling that the author acknowledges the use of the term but is keen to distance themselves from any naïve (or worse) views associated with it. The reason, of course, is the existence and effects of racism, and we shall have more to say about that below. Initially, though, we will concentrate on what a narrow biological approach might have to say about human races.

We can begin by noting that the meaning of race when applied to humans has shifted over time. The word emerged in the sixteenth century in English from the Italian *razza* and the French *race* and referred to a group of organisms (whether humans or otherwise) connected by common descent (Barnshaw 2008). Over the next two centuries, this criterion was increasingly joined by one to do with physical similarities, such as in skin color, hair texture, and facial features. If we fast forward to today, the use of the term "race" as applied to humans has largely disappeared within science. The principal reason for this is the way the term has been abused, but there are a number of reasons to do with objective science too.

For a start, there is no agreement as to how humans (who all belong to one interbreeding species, *Homo sapiens*) should be divided into races. Now, this isn't a knockdown argument—we have already seen that when applied to any species, the notion of races is informal with a degree of subjectivity. Nevertheless, the suspicion remains that the classification of humans into different races tells us more about the person who is doing the classifying than it does about the races themselves.

Indeed, in the first half of the twentieth century, reacting against what he felt was spurious science being used to identify discrete human races, the anthropologist Franz Boas (1858–1942) presented statistically solid arguments to substantiate his conclusion that phenotypic variation within so-called races is greater than the average between them and that skin color is not strongly associated with clusters of other traits (Boas 1928, 1940; Jackson & Depew 2017). He also showed that head shape and size—held by many anthropologists of the day to be traits that could reliably be used to categorize individuals to different races—were variable, depending on environmental factors, particularly nutrition and health. No need here for Boas to appeal to organicism or holism. Even a narrow, reductionist reading of the data—so long as such a reading was objective and careful—showed that the standard views of the day about races lacked validity.

Boas's pioneering work has been corroborated by subsequent biological research. He also argued that differences between people in respect of their behavior are not due to innate differences but are the result of upbringing and culture. Finally, Boas introduced the notion of "cultural relativism"—the idea that different cultures (he also used the word "civilizations") and cultural practices cannot be ranked as superior or inferior. (Funeral practices are a good example—cultures simply vary as to whether the corpse is bathed, touched, buried, cremated, mummified, left to decompose or be scavenged, subsequently dug up, and so on.) Rather, all of us see the world through the eyes of our own culture. One of the great advantages of traveling, or even just reading about other cultures, is that it can help us to see through different eyes too, often making us less sure that our culture ("the ways things are done round here") is *the* right way of doing things.

As it became possible in the second half of the twentieth century to use reasonably objective measures to determine the differences between people at the molecular level (e.g., through the analysis of proteins or genes), it was increasingly found that while it was possible statistically to see differences between the *averages* of these measures between people assigned to different races, the measures themselves were not much use at assigning individuals to these different races. It's rather as if we used height to distinguish men and women. Yes, on average, men are taller than women, but we all know men who are shorter than the average woman and women who are taller than the average man. In other words, height, on its own, is not a very useful predictor of sex. In fact, humans turn out to have surprisingly little genetic variation compared to most mammal species, including our closest relatives. Furthermore, Richard Lewontin (1972), whom we cite on a number of occasions in this book, showed that very little of what genetic variability there is can be explained by assigning us to different races—a finding backed up by subsequent research (Rosenberg et al. 2002; Goodman et al. 2019).

In the previous chapter, we saw that classifying humans as males or females is not as straightforward, possible, or useful as is often supposed. Nevertheless, it is the case that many features shared by most women cluster, as do many features shared by most men. To state the obvious, this is why most of us on meeting someone for the first time almost instantaneously classify them as male or female and are not surprised that in the large majority of cases, other people, including the person we have just classified, reach the same conclusions. However, this is much less often the case with race. Again, the issue is to do with individuals and populations.

Given enough biological data, one can begin to establish average differences between "races," but if one tries to assign individuals to such races, one is more likely to find disagreement with others, including the person themself. It is important not to overstate this point. Much writing on race claims too much, asserting that the division of people into races has absolutely no scientific validity. The reality is that people do differ in their anatomy in ways that are related to their ancestry. Groups of people who have lived for many generations near the poles nearly always have paler skin than do groups of people who have lived for many generations in the tropics, and the reason for this that many of us were told at school—that dark skin is advantageous at reducing the incidence of certain skin cancers but disadvantageous in that it reduces the rate at which the body can make vitamin D, which we require—is still believed to be true (University of California Museum of Paleontology 2019).

Nevertheless, there are a number of biological reasons why it's less valid to assign humans than many other species to races. For a start, we are a relatively "young" species. *Homo sapiens* has probably been around for about 200,000–300,000 years (Hublin et al. 2017). That may sound a long time, but the typical mammalian species lasts for about one million years. Then there is the fact that it is very likely that all humans originated in Africa—the "Out of Africa" hypothesis (Wilshaw 2019). What this means is that until perhaps 70,000 years ago, all of our ancestors would have been assigned to a single race (if the criteria nowadays used by some to identify human races had been used); there simply hasn't been a great deal of time in evolutionary terms for distinct races to evolve. Furthermore, humans have always moved around a lot, and that is increasingly the case nowadays. Such movement leads to the exchange of genetic material between populations and militates against consistent genetic differences existing between different groups.

RACISM

Assigning people to different races might have been a relatively innocuous pastime, even seen as akin to cataloguing different varieties of roses, were it not for a long history of misuse of such practices. Racism entails prejudice, discrimination, or violence against people because they are seen to be of a different race or ethnicity. There is, of course, a certain irony in such a definition. It is racists who place the most weight on the idea that races exist in an objective sense. Yet, those who experience racism and those who campaign against it find that they need to make some use of the term if only to show the extent of the harms that may be experienced as a result of a person's race.

Racism gets compounded with tribalism, ethnicity, and nationalism. It has probably been around long before the advent of recorded history. There is certainly plenty of evidence of it in the Hebrew Bible—see how the Jewish people were treated in Egypt under Pharaoh in the generations after Jacob (Israel) and his twelve sons and their families (some seventy in all) moved there from Canaan during the years of famine, and how, after the Exodus, some 430 years later, the Jewish people set about slaughtering the indigenous people (the Amalekites, the Canaanites, and so forth) as they sought to find a land for themselves. Later, as the Jewish people returned from the exile in Babylon, the books of Ezra and Nehemiah recount how the existing prohibition on intermarriage was rigidly enforced:

> 9 Then all the men of Judah and Benjamin assembled at Jerusalem within the three days; it was the ninth month, on the twentieth day of the month. And all the people sat in the open square before the house of God, trembling because of this matter and because of the heavy rain. 10 And Ezra the priest stood up and said to them, "You have trespassed and married foreign women, and so increased the guilt of Israel. 11 Now then make confession to the Lord the God of your fathers, and do his will; separate yourselves from the peoples of the land and from the foreign wives." 12 Then all the assembly answered with a loud voice, "It is so; we must do as you have said. 13 But the people are many, and it is a time of heavy rain; we cannot stand in the open. Nor is this a work for one day or for two; for we have greatly transgressed in this matter. 14 Let our officials stand for the whole assembly; let all in our cities who have taken foreign wives come at appointed times, and with them the elders and judges of every city, till the fierce wrath of our God over this matter be averted from us." 15 Only Jonathan the son of As′ahel and Jahzei′ah the son of Tikvah opposed this, and Meshul′lam and Shab′bethai the Levite supported them. (Ezr. 10:9–15)

The distinction between race and ethnicity is not clear-cut, but race is nowadays used more often in relation to physical characteristics, notably skin color but also such features as hair texture, whereas ethnicity draws more heavily on national, tribal, religious, linguistic, or cultural origins or backgrounds (Blakemore 2019). Most of us are used to being given a choice of possible races from which to choose when describing ourselves on some forms: White, Black or African American, Asian, American Indian, Alaska Native, Native Hawaiian, and Other Pacific Islander is a common list in the United States. In the United States, rather oddly, the official Census has only two possibilities for ethnicity: "Hispanic or Latino" and "Not Hispanic or Latino." In the UK, the current official government advice is that if information on ethnicity is sought, people are asked the following question with the following possible responses:

WHAT IS YOUR ETHNIC GROUP?

Choose one option that best describes your ethnic group or background

White

1. English / Welsh / Scottish / Northern Irish / British
2. Irish
3. Gypsy or Irish Traveller
4. Any other White background, please describe

Mixed / Multiple ethnic groups

5. White and Black Caribbean
6. White and Black African
7. White and Asian
8. Any other Mixed / Multiple ethnic background, please describe

Asian / Asian British

9. Indian
10. Pakistani
11. Bangladeshi
12. Chinese
13. Any other Asian background, please describe

Black / African / Caribbean / Black British

14. African
15. Caribbean
16. Any other Black / African / Caribbean background, please describe

Other ethnic group

17. Arab
18. Any other ethnic group, please describe

(Office for National Statistics 2021)

Racism is perhaps especially associated nowadays in people's minds with the treatment of indigenous and Black people by White people over the last few hundred years. The reasons behind such treatment are primarily to do with Renaissance and post-Renaissance European military superiority that exacerbated the existing institution of slavery and led to many countries being colonized by the Belgians, the British, the Dutch, the French, the Germans, the Italians, and the Portuguese, so that to this day, huge swathes of the globe still live with a colonial legacy. The so-called Scramble for Africa is illustrative of the extent of colonialism. In 1870, some 10 percent of Africa was under European control; by 1914, the figure was close to 90 percent.

In the case of South America, some fifty-six million indigenous people are thought to have died as a result of such European diseases as measles, smallpox, influenza, and the bubonic plague between Christopher Columbus's arrival in 1492 in the Caribbean and the early 1600s (Koch et al. 2019). That is 90 percent of the pre-Columbian indigenous population. The population loss meant that previously farmed areas returned to their natural states. This is thought to have led to such an increase in photosynthesis that the resulting decrease in atmospheric carbon dioxide levels contributed to the Little Ice Age of the 1600s as global temperatures cooled.

The Conquistadors may not have meant to exterminate countless South American tribes through the diseases they brought with them, but many national acts of racism have been entirely intended. Australians, for example, are still coming to terms with the history of the treatment of Aboriginal and Torres Strait Islander men and women, the descendants of those who had originally colonized the continent some 65,000 years ago. British colonists "dispersed" indigenous peoples from their lands by either moving or massacring them, and women were routinely sexually exploited (C. Thomas 1991). To this day, Aboriginal and Torres Strait Islander children have substantially poorer educational outcomes (Department of the Prime Minister and Cabinet 2017), while life expectancy for Aboriginal and Torres Strait Islander people is about a decade less than for nonindigenous people (Australian Institute of Health and Welfare 2020).

We could continue in this vein for many pages, but more directly relevant to our subject is the role that science has played in people's attitudes toward and treatment of those they consider different from themselves.

Scientific Racism

Science has long played a key role in racism. With its claim to be the source of objective, empirical knowledge, science has contributed to racism in two main ways. First, as we have seen above, despite a long legacy of the rejection of the scientific basis for assigning people to races, science has often not just acquiesced in such categorizations but also led the way. Second, and perhaps more importantly, many scientists have helped build up a story in which races are arranged hierarchically, with White people at the top (Sussman 2014). Before we get into an account of this, it is worth emphasizing that to arrange races hierarchically is to fail to understand how evolution works. Darwin himself back in 1837 (when still in his twenties—he was born in 1809—and just returned from his five-year voyage on *The Beagle,* and more than twenty years before his *On the Origin of Species* finally appeared in 1859) had written in his notebook: "It is absurd to talk of one animal being higher than another. We consider those, when the cerebral structure / intellectual faculties most developed, as highest. A bee doubtless would when the instincts were" (Darwin 1987, B: 74).

Evolution is all about adaptation to the prevailing environment. Most of us (some parasitologists excepted) may admire cheetahs and kingfishers more than their intestinal parasites, but from evolution's perspective, each has provided a perfectly successful set of descendants from LUCA (the Last Universal Common Ancestor) some 3.9 or more billion years ago (Betts et al. 2018). And the same point holds about races within a species. Either the small differences between races are of no adaptive significance, being simply the result of such factors as genetic drift, or races are adapted to local environments. In the latter case, there are no higher or lower races; of course, it would make sense to talk about one race being better adapted than another in a particular environment, just as thermophilic bacteria are better adapted at living in hot springs, and sloths in tropical rain forests.

The father of modern taxonomy was Carl von Linné (1707–1778), the Swedish biologist who is better known by the Latinized version of his name: Carolus Linnaeus. Linnaeus introduced the system of binomial nomenclature, in which all species have two Latin words—a genus (e.g., *Felis*—small and medium-sized cats) and a specific epithet (so the European wildcat is *Felis silvestris,* the African wildcat is *Felis lybica,* and so on). When it came to humans, Linnaeus classified us as all belonging to one species, which he named *Homo sapiens,* but in the 1758 and most important (the tenth) edition of his groundbreaking *Systema Naturae,* he divided us into six subspecies (Marks 2007): Wild Man, American, European, Asiatic, African, and a catch-all category that included Patagonians, Hottentots, Chinese, and others.

Some of the features that Linnaeus used to distinguish his subspecies, while unacceptably stereotypical nowadays (Europeans have yellow or brown hair and blue eyes; Africans have black, frizzled hair, flat noses, and tumid lips), were attempts at relatively neutral anatomical descriptions. The greater problem comes with some of the other attributes listed. While Linnaeus didn't arrange his human subspecies into a hierarchy, he clearly thought very differently of them. Americans were said to be governed by customs, Europeans by laws, Asiatics by opinions, and Africans by caprice.

Linnaeus was a European, and the overwhelming majority of recorded science in the eighteenth and nineteenth centuries was undertaken and published by Europeans. It is not surprising, therefore, that however the classification of humans was done (subspecies or races, Europeans or

Caucasians), White people came out on top. This became even more evident once biologists began to arrange human races hierarchically.

The scientist who in the last fifty years perhaps did the most to bring scientific racism to public attention was the paleontologist and brilliant writer of natural history Stephen Jay Gould in his *The Mismeasure of Man* (1981). Gould pointed out that an important post-Linnean step was taken by the German anthropologist Johann Friedrich Blumenbach (1752–1840). As far as human races go, Blumenbach did two main things. First, he divided humanity into the five races whose names pretty much survive to this day: Caucasians (people of European, Middle Eastern, and North African origin); Mongolians (all East Asians and some Central Asians); Malayans (Southeast Asian and Pacific Islanders); Ethiopians (sub-Saharan Africans); and Americans (American Indians). Second, and perhaps more importantly, he argued that of the five races, Caucasians were the most beautiful.

Nowadays, it seems obvious to us that this is about as subjective a criterion you could come up with (there is an objective side to beauty—for example, symmetrical faces are nearly always identified by people as being more beautiful than asymmetrical faces, but this has nothing to do with interracial differences). However, it helped set a very important and unpleasant train in motion. This is all rather ironic, since, as Gould pointed out, Blumenbach can hardly be straightforwardly characterized as racist. Blumenbach argued strongly that all humans were related by descent (monogenesis), against the view that races had been created separately (polygenesis); he pointed out that races were not discrete and believed that racial differences would be eliminated if people moved to another part of the globe; he campaigned for the abolition of slavery; he argued in favor of the moral and intellectual unity of all people; and he had a special library in his house devoted exclusively to writings by Black authors. Indeed, Gould (1981/1996, 405) concluded that "Blumenbach was the least racist, most egalitarian, and most genial of all Enlightenment writers on the subject of human diversity."

There are plenty of examples of other scientists who felt, *contra* Blumenbach, that races of people could be distinguished by their moral and intellectual capacity. Indeed, the issue of race and intelligence regularly resurfaces and exists to this day. Before we even get onto race, there are those who are deeply suspicious of attempts to measure intelligence, arguing that such attempts are inevitably culturally biased or don't tell us much about what good education could do to improve a person's score on an intelligence test. Then there are those who argue that genetics plays no part in a person's intelligence, although this argument is getting harder to substantiate, as genetics plays a part in just about any human characteristic—the problem is that many people equate genetics with determinism and so presume that if genetics is involved in something, that something is "determined" in some sense. Finally, we get on to the relationship between race and intelligence.

In 1969, Arthur Jensen published as article in the top-ranking journal *Harvard Educational Review* (Jensen 1969). His article has the title "How much can we boost IQ and scholastic achievement?" and his short answer (the article is 123 pages long) is "Not much." Jensen argued that what he terms "compensatory education" is a failure and cited other authors who had studied interracial differences in intelligence (measured using intelligence quotient [IQ] tests), concluding, for example, that "on the average, Negroes test about 1 standard deviation (15 IQ points) below the average of the white population in IQ" (81). Unsurprisingly, Jensen's article proved deeply controversial and was heavily critiqued as soon as it was published. It gave rise to many studies that reached opposite conclusions. For example, Schiff et al. (1982) undertook a study in France where thirty-two children abandoned at birth, whose mothers and fathers were all unskilled workers, were adopted at about four months of age into families with fathers within the top 13 percent of the socio-professional scale. The IQs of these thirty-two children were measured in school and compared with those of two other sets of children of comparable age: those in existing large-scale studies on children of unskilled workers, and twenty half-siblings of the thirty-two but who had been raised by their biological parents rather than adopted. It was found that the adopted children had IQs that were on average fourteen points higher than the children in these two other sets. These results suggest that irrespective of whether compensatory education is a failure, the home environment in which children grow up plays a major part in their intellectual development. They also show the way in which science can be used to undermine the purported intellectual bases of racism, just as Boas had done decades earlier.

Such refutations of Jensen's conclusions about the pointlessness of trying to compensate for the poor performance of certain groups on IQ tests have not caused Jensen's ideas to be abandoned. These ideas tap into widespread racist views and regularly resurface. A particularly influential example of this was provided by the publication in 1994 of the best-selling *The Bell Curve* (Herrnstein & Murray 1994). To be fair to the authors, their conclusions on the relationship between race and intelligence were less clear-cut than those of Jensen (1969), and they argued that while both genes and the environment were implicated in interracial differences in intelligence, the relative contributions could not be determined. More recently, Murray (2020) has revisited this and related issues in a book titled *Human Diversity: The Biology of Gender, Race, and Class.* He hasn't softened his views, writing, for example, "racism and sexism still play a role in determining who rises to the top. But that role is not decisive" (205) and, imagining a reader who is not persuaded by his arguments, "defending your belief that ethnic differences in IQ are meaningless is tough" (206).

The ongoing sensitivities around the issue of race and intelligence are indicated by the reception given to the recent publication of an article by an Oxford doctoral student, Nathan Cofnas (2020), in which he argued that the widespread academic practice of ignoring or rejecting research on intelligence differences between groups, including racial groups, is a mistake. The editors of the journal in which he published were well aware of these sensitivities and, in an accompanying editorial note wrote:

> The decision to publish an article in Philosophical Psychology is based on criteria of philosophical and scientific merit, rather than ideological conformity. As a prime laboratory of blue-sky thinking, the philosophical debate is the forum par excellence for questioning consensus beliefs. Ideological red lines may still exist, but they are drawn more widely in this context than in, for example, the political forum. This principle is put to the test when the ideological undertones of a paper (and no article is free of them) do not square with one's own sensitivities. (van Leeuwen & Herschbach 2020, 148)

This attempt at calming the situation did not work. A petition was launched, and it attracted more than 120 signatures. It described the article as "disingenuous," stated that it was not "competently reviewed," and concluded:

> For this reason, we call upon the leadership of *Philosophical Psychology* to respond. Potentially responses include apology, retraction, or resignation (or some combination of these three). Should they choose to resign, we demand that a new group of leaders openly and honestly articulate a plan to reform the peer-review practices of the journal. Until the leadership respond in an acceptable way, we call upon philosophers and other researchers to boycott the journal by refusing to submit papers to it or referee for it. (Alfano 2020)

Genetic Ancestry Testing

Many people are fascinated by their "roots"—where they and their ancestors came from. Traditionally, knowledge of this relied on family lore, public records (birth certificates, marriage certificates, wills, and so on), and visiting churchyards or other places where people were buried. This rather gentle pursuit has been turned on its head in recent years by the arrival of genetic ancestry testing. Firms such as 23andMe and AncestryDNA use DNA testing to look at specific places on a person's chromosomes to determine genealogical relationships or estimate the ethnic mix of the person. This approach to understanding oneself is about as reductionist as one can get.

Forensic DNA testing works on much the same principles and has been used for some three decades in criminal investigations, although even here there is still sometimes a naïve presumption that DNA can prove who the criminal was, whereas the most that it can do is identify (to a very high degree of accuracy) from whom biological samples come—how they got to be where they are requires traditional investigative work. More recently, there has been an explosion of interest in genetic ancestry testing among the general public. At first sight, the promise is alluring. For a relatively modest fee, you can be told the different places your ancestors came from and if there are others in the company's database whom you are likely to be related to and how closely. In reality, though, it's not quite as straightforward as this. For a start, how far back do you want to go? If you only want to go back one generation, most of us (but not all of us, of course) know where our mother and father lived. Go back 70,000 years or so and we all come from Africa.

That aside, there are two main ways in which genetic ancestry testing may not live up to the expectations of some of those using such services. The first is to do with the accuracy of the "findings" (e.g., Walajahi et al. 2019); the second to do with how they are "heard." There are also potential concerns about how the information might be used in some countries—might police forces or governments be able to gain access to the genetic information that results from people sending in their tissue samples? And, of course, you have just handed over samples of yourself to a private company for DNA sequencing.

Most of the companies that operate in this field divide the world up into different geographical regions and tell you what percentage of your DNA comes from each region. Of course, we don't have genetic records from hundreds of years ago. So, what these tests are actually doing is telling you not, for example, that 20 percent of your ancestors at some time in the distant past lived in Scandinavia but rather that 20 percent of your DNA is typical of people who nowadays live in Scandinavia and are not known recently to have moved there from elsewhere.

For an enthusiastic advocate of such testing, see Mercedes (2021), who not only found out, to her surprise, that she was 43 percent Germanic Europe; 21 percent Eastern Europe and Russia; 13 percent England, Wales, and Northwestern Europe; 10 percent Norway, and so on, but also that she probably had ancestors among the colonists who lived in New England, and that as of January 29, 2019, she had 496 DNA matches at fourth cousin distance or closer, with new DNA matches coming in "almost every day!"

There is little doubt that commercial ancestry testing is improving in accuracy, and it might be supposed that inaccuracies do not matter too much for what is often a recreational activity. Nevertheless, some people take the result of the tests very seriously. In-depth interviews suggest that people often "hear" the results of genetic ancestry testing selectively, using it to reinforce narratives about themselves that they prefer and dismiss narratives that don't fit with their perceptions of themselves. For example:

> Shannon, a 57-year-old artist, had been adopted and thought of herself as white until, at age 18, her adoptive father told her that she had Native American ancestry. After that, she strongly identified as Native American as well as white. She took an admixture test to confirm her Native ancestry, but it came back reporting that she had none. She declared: "The first DNA [test] we did was to find out our . . . ancestry, how much Indian and stuff and when that came back, I about fell over because mine said zero for Native American. And I'm like 'okay, either my dad isn't my dad or you know, what is going on here?' I mean I was literally hysterical, that's how much it means to me to be Indian." While Shannon claimed she did not fully understand how the test worked, she decided that it simply must be wrong and continued to identify as white and Native American. (Roth & Ivemark 2018, 166)

Systemic Racism

It is clear from the account in this chapter to date that while the term "race" has its roots in conventional biology, a reductionist approach that looks only at intergroup genetic differences is utterly insufficient, on its own, to understand the issue of race as applied to humans. The advantages of looking at race and racism more holistically become even clearer when we look at the phenomenon of systemic racism. Systemic racism is sometimes called "structural racism" or "institutional racism," but we use the term "systemic racism" here, as it draws attention to the ideas of systems biology, in which the level of analysis is the system as a whole (e.g., an entire cell or an ecosystem) rather than focusing only on its separated components.

The central point about systemic racism is that the injustices of racism result not only from the attitudes and actions (intentional or otherwise) of individuals but also from discriminatory procedures and practices that are embedded in an organization or the broader society. Systemic racism reaches its apogee in systems of apartheid, notoriously that in South Africa and what is now Namibia from 1948 to the early 1990s, where notions of White supremacy and an entire legal system and associated set of cultural practices led to political, social, and economical domination of the majority Black population by the country's minority White population (Mayne 1999).

But systemic racism is found worldwide and in ways that are not always obvious. An example of how it operates is provided by the actions of the Home Owners' Loan Corporation, established in 1933 in the United States as part of Roosevelt's New Deal. Its primary aim was to provide refinancing for home mortgages that were in default to prevent foreclosures. Borrowers gained, as they were offered loans at lower rates (5 percent rather than the 6–8 percent typical at the time) and over longer periods (fifteen years rather than the three to twelve years that was then customary), thus reducing their regular payments. All sounds good? However, lending money is a risky business, and some 20 percent of loans were never repaid. To protect its interests, the Home Owners' Loan Corporation introduced the practice of "redlining" neighborhoods that were calculated to be at high risk of defaulting on home loans. Individuals in such neighborhoods, who were more likely to be African Americans, were therefore denied the possibility of obtaining such a loan. A recent study has shown that cities that the Home Owners' Loan Corporation categorized in this way became increasingly racially segregated compared to those it did not (Faber 2020). These effects persist to this day and helped fuel the move of White middle-class Americans to the suburbs while African Americans were more likely to remain in city centers that became increasingly run down. Positive feedback set in, making it very difficult to reverse such trends.

A contemporary example of systemic racism is provided by Aaron Harvey's account of how, as a young Black man raised in the historically Black community of Lincoln Park in San Diego, he had had more than fifty encounters with local police, despite having no criminal record:

> In the summer of 2014, I was one of over a dozen Black men from my community who were picked up and thrown into jail on charges of conspiracy to commit nine shootings I knew nothing about. It was then that I found out that the San Diego Police Department had designated me as a member of a local gang when I was 18 years old.
>
> Under California's gang conspiracy law, anyone designated as a gang member can be held liable for a crime allegedly committed by the gang, even if they had no knowledge of the crime. Individuals who have been designated as being gang members are also subject to harsher sentences through enhancements.

> Despite having no knowledge of the crimes I was charged for—a fact District Attorney Bonnie Dumanis acknowledged in court—I was facing a sentence of 56-years-to-life because my name was in the CalGang database.
>
> From jail, we fought our case for seven months. During my trial, the detective investigating the unresolved shootings testified that he had the legal right to charge and incarcerate all 543 individuals in my neighborhood who'd been designated as being gang members.
>
> The judge in the case eventually dismissed the charges, but those seven months of being locked up turned my life upside down. I was ripped away from my loved ones, I lost my home, my education was cut short, and I was financially ruined. All for being on a list I knew nothing about. (A. Harvey 2016)

A subsequent audit revealed that nearly 20 percent of the people in the CalGang database were African American and 66 percent Latino, while only 7 percent of Californians are African Americans and just 38 percent are Latino. Furthermore, forty-two individuals *under one year of age* had been added to the CalGang database.

Systemic racism is being exacerbated by the way in which some artificial intelligence (AI) algorithms work (Benjamin 2019). For example, if an algorithm overtly and explicitly favors US rural over US urban individuals, that can be a form of racial coding, where rural stands as a proxy for White and urban for Black. The way that AI algorithms work is such that such equations can be made both in a way that is more difficult to unpick and in a way that some are more likely to consider "objective" and free from bias (Burbidge et al. 2020).

New technologies can exhibit racist bias for all sorts of reasons. Benjamin (2019) tells the story of how when Princeton University media specialist Allison Bland was driving through Brooklyn, the Google Maps narrator told her to "turn right on Malcom Ten Boulevard." As Bland tweeted, "I knew there were no black engineers working there." But there are more insidious instances. Angela Helm (2016) gives the example:

> Type "three black teenagers" into a Google image search, and a bunch of menacing police mug shots turn up. "Three white teenagers" results in smiling young folks who look to be selling Bibles door-to-door.

What was going on here was not that Google had designed their search algorithms to come up with such instances of racial profiling but that the algorithm reflected typical (average) user searches—search algorithms work by make good guesses as to what we want to find, rather like predictive texting. At the time, a Google spokesperson said:

> Our image search results are a reflection of content from across the web, including the frequency with which types of images appear and the way they are described online. This means that sometimes unpleasant portrayals of sensitive subject matter online can affect what image search results appear for a given query. These results don't reflect Google's own opinions or beliefs. As a company, we strongly value a diversity of perspectives, ideas and cultures. (Sini 2016)

We are sure you feel better for that.

Benjamin (2019) points out that some technologies fail to see Blackness, while others render Black people hypervisible and so expose them to systems of racial surveillance. An example of how a technological practice can do disservice to Black people is provided by so-called Shirley cards. These are named after Shirley Page who worked at Kodak and whose photograph was used as a way of getting the color contrast as good as it could be on photographs. She eventually got married and left the company, but the new cards with new photographs continued to be called Shirley cards. There was something else the cards had in common—the women were always White. For more than twenty years, this meant that White was literally the norm. As a result, Black, it was said, just didn't photograph as well.

In an ongoing project, Lorna Roth (2021) has shown that it wasn't the many complaints from Black people, including parents whose sons' and daughters' graduation photographs hadn't come out as they had hoped, that led to Kodak changing. It was complaints from furniture and chocolate manufacturers (you couldn't make this up). Chocolate companies were saying that they "weren't getting the right brown tones on the chocolates" in the photographs (Lewis 2019). Kodak finally responded by producing more diverse Shirley cards and produced a new film, Kodak Gold Max, advertised as being "able to photograph the details of a dark horse in low light"—a coded message for being able to photograph people of color rather better than had previously been the case with its film.

It might be supposed that the issue has gone away with the demise of film and the rise of digital photography, but this is not the case (Lewis 2019). Those with darker skin are often disadvantaged when photographed, and facial recognition software typically does a worse job for those of color; darker-skinned females are the least well served (Najibi 2020). In a further instance of systemic racism, cameras that are used to take photographs that are then analyzed using facial recognition software are typically not geographically evenly installed. In Detroit, Project Green Light was enacted in 2016, installing high-definition cameras across the city. However, the cameras were far more likely to be placed in majority-Black areas than in predominantly White or Asian areas (Najibi 2020).

THE PHENOMENON OF RACE

So, how are we to understand race? A variety of positions can be taken (Glasgow et al. 2019), but the most common academic position nowadays is to see it as a social construction. Genetics, physiology, anatomy, and overall morphology all play a part in who each of us is and in how we understand ourselves—as we also saw in Chapter 5 in relation to sex—but they do not define us or our worth. Understanding race as a social construction allows it to be seen both as a sociopolitical reality—which it most certainly seems to be—and as a way that individuals can choose to self-identify, without fossilizing any of us into rigid categories. Indeed, it seems likely that the descriptions we choose to identify our race / ethnicity will continue to develop over time, rather as the country to which we belong may be of great importance to us, for all that we acknowledge that countries, both in their borders and in their names, have often been, and no doubt will continue to be, subject to change.

Interestingly—and encouragingly—both a mechanistic / reductionist and an organismic / holistic understanding of race provide no support for racism. Either way of understanding biological reality fails to support an argument that positions some people as inferior and others as superior on the grounds of race. We find ourselves in agreement with Adam Rutherford:

> Of course, racism is not simply wrong because it is based on scientifically specious ideas. Racism is wrong because it is an affront to human dignity. The rights of people and the respect that people are due by dint of being a person are not predicated on biology. They are human rights. Hypothetically, if there were genetic differences between populations that we have not found yet, and these do correspond with the folk definitions of race, the fact that we have not found them means that they are tiny at best. If those things were true—and there is no evidence that they are—would that have any impact on how we should treat each other? (Rutherford 2020/2021, 159)

Chapter Seven

MOTHER EARTH

HOLISTIC HARVEST

I'm farmer Henry Gwynn, a seventh generation Tallahassee native, and with the help of my hard-working fiancée Burgen Schwartz, run a permaculture farm just south of Tallahassee, Treehouse Permaculture. When selling produce at the farmers market, I am often asked: "What is permaculture and how does it compare to conventional farming?" Permaculture is a sustainable design science, a way of integrating indigenous agricultural practices into modern food production.

At our farm we chose practices that enrich soil life, plant diversity, and the local food community. Most conventional agricultural models do not establish or restore these vital systems. Whether it is stripping soil life for faster profits, contributing to fertilizer run-off in our waterways, or disturbing essential wildlife corridors, conventional farming practices don't consider the bigger, long-term picture. Our core intention is to design our farm with a focus on sustainable food production and habitat restoration.

Our holistic land management practices also include goat grazing. Our goats clear fire suppressed thickets not only to provide fertile growing space, but to restore the land to a healthy longleaf pine savannah. Now native wildflowers and wild fruit are coming up in these pastures.

Our methods and many permaculture ideas can help you live closer to nature and your food. Whether you live in a city, have a farm, or homestead, these ideas can be applied at any scale to help humanity have a sustainable future.

—GWYNN 2020

THE WORLD AS A MACHINE

This article from the *Tallahassee Democrat,* August 24, 2020, sets the topic and theme for this chapter. Today, there is a vocal group—a very vocal group, and not just in California—claiming that society has made a disastrously huge mess of the world in which we live. We have worshipped at the false god of "mechanization" and failed to follow the true path to the celestial city, the path of "holism." This is a question not just of facts but of morals. We have destroyed our home—"Mother Earth"—making for an inadequate life today with the prospects of even worse lives for our descendants. Before it is too late, we must turn from the machine metaphor and embrace the organism metaphor.

Are these critics right? Are they exaggerating, pumped-up doomsday prophets in organic sandals making mountains from molehills? And how did they come to this conclusion? As always, we turn first to the past and then move to the present.

We return to Plato, for it was he who saw the world, the whole universe, as an organism, something good in itself, ever giving and sustaining. Picked up by the Christian philosophers and theologians, it all made good sense of their world picture. The world as organism did not, of course, exist in its own right. It was the creation of our good God. For this reason, it would be wrong to worship it, as do the pagans. But it is not just dead matter; it is a living thing—our mother! Then came the Scientific Revolution and the all-conquering machine metaphor. The early sixteenth-century humanist and physician, often known as the "father of mineralogy," Georg Agricola (1495–1555), put his finger right on the way that the organic metaphor was standing in the way of progress, meaning the machine metaphor. As machines get ever more sophisticated, you move from making them entirely from organic materials—wood specifically—and turn more and more to minerals. Without metal, you simply could not make a decent clock. The trouble is that the organicists could and would argue that obtaining these minerals, mining, violates the Earth. First, Agricola gave the traditional argument:

> The earth does not conceal and remove from our eyes those things which are useful and necessary to mankind, but on the contrary, like a beneficent

> and kindly mother she yields in large abundance from her bounty and brings into the light of day the herbs, vegetables, grains, and fruits, and the trees. The minerals on the other hand she buries far beneath in the depth of the ground; therefore, they should not be sought. But they are dug out by wicked men who, as the poets say, "are the products of the Iron Age." (Agricola [1556] 1950, 6–7; the translation is by the future president of the United States, Herbert Hoover, who was a mining engineer at the time.)

Then, he went after the argument:

> If we remove metals from the service of man, all methods of protecting and sustaining health and more carefully preserving the course of life are done away with. If there were no metals, men would pass a horrible and wretched existence in the midst of wild beasts; they would return to the acorns and fruits and berries of the forest. They would feed upon the herbs and roots which they plucked up with their nails. They would dig out caves in which to lie down at night, and by day they would rove in the woods and plains at random like beasts, and inasmuch as this condition is utterly unworthy of humanity, with its splendid and glorious natural endowment, will anyone be so foolish or obstinate as not to allow that metals are necessary for food and clothing and that they tend to preserve life? (14)

Note that the argument is as much moral as scientific or anything else. On the one side, the claim is that the world in itself has value, and to treat it in certain ways is a violation of this thing of value. On the other side, the claim is that not to use our knowledge and abilities to further our well-being is no less immoral. We shall see that this divide continues.

After the Scientific Revolution, mineralogy specifically and geology more generally became ever more needed (Rudwick 1972, 2005). Hugely significant was the search for new sources of energy to drive the move to mechanization and industrialism. Coal become all important, as did more efficient methods of mining. Obviously, it was not just a matter of getting coal out of the ground. You needed ways of getting it to factories, and no less did you need ways of getting your finished products to their destinations. Moving things on water rather than by road had long been seen as far more efficient. Supplementing rivers and (especially pertinent for an island such as Great Britain) the seas, canals soon started to crisscross the lands. This brought on another need of geological expertise. If one is faced by a mountain, is it a better strategy to excavate a tunnel through the mountain or to build a canal, with an elaborate series of locks, going over the top of the mountain? The first option makes sense if the mountain is made of limestone or sandstone. If it is solid granite, better go over the top.

GEOLOGICAL THEORIES

By the eighteenth century, there were two rival secular theories of the Earth. First was Neptunism, from Germany, the child of the Freiberg mining engineer, Abraham Gotlob Werner. It saw water and weathering as the key causal geological phenomena. At first, the globe was covered in water, with many minerals held in suspension. Then, they started to descend and form the strata covering the Earth, with heavier ones such as granite toward the bottom and lighter ones such as limestone toward the top. Fossils attest to the various organisms living at the time of deposition. As time went by, the earth buckled and pushed, and land appeared above the surface, and then we got weathering, producing the Earth's crust as we now have it. A later version of this theory was produced at the beginning of the nineteenth century by Georges Cuvier ([1813] 1822). Informed by his fossil discoveries, he saw periodic revolutions on the Earth's surface, with widespread flooding, entailing the extinction of present forms and then the arrival from elsewhere on the globe of new forms (Coleman 1964). He was no evolutionist nor was he keen on the idea of biological progress. Any appearance of this was a chimera.

Second was Vulcanism. This was the creation of the Scottish physician and farmer, James Hutton (Hutton 1788; Repcheck 2003). It argued that the center of our planet is hot liquid rock. The surface pressure from this and its occasional breaking through and bubbling out is the chief causal factor in making Earth geology. Hutton was not indifferent to weathering and erosion, but it came secondary. Apparently, everything goes in continual cycles, with the hot liquid bubbling up and solidifying. It then pushes across the surface and finally is forced down below again. Here, it is melted down and everything begins all over again. Neptunism is generous with time—no one was going to be constrained to the biblical 6,000 years since the creation—but nothing to Vulcanism. It saw an infinite span. Hutton is famous for his saying, "We find no vestige of a beginning, no prospect of an end" (Hutton 1788, 304).

Neither approach was especially favorable to the organic metaphor, for all that in the nineteenth century (especially in Britain), there were those who saw in the Cuvier-type revolutions a place for divine intervention—particularly as He created new species to replace those swept away by the floods and the like. Increasingly, however, this "Catastrophism" gave way to an updated form of Vulcanism, "Uniformitarianism," developed and promoted by the Scottish lawyer-turned-geologist Charles Lyell (1830–1833). Then, in the twentieth century, this evolved into its successor, continental drift—continents moving around the globe—fueled by plate tectonics, where liquid rock comes up from below forcing endless cycles as it solidifies and then is forced back underground (Hallam 1973; Ruse 1981). One hardly needs to draw attention to the extent to which plate tectonics harks back to Vulcanism. Likewise, one hardly needs to draw attention either to the extent to which Vulcanism—and plate tectonics in its turn—reflects the machine model of modern science. One of the greatest inventions of the eighteenth century, for pumping water out of mines, was the Newcomen engine. It works by producing steam, which then goes in a cycle of condensation and reheating, as it drives a pump (Figure 7.1). It is a man-made counterpart to the circular mechanism driving and creating the Earth in the Vulcanist picture of things.

The machine metaphor has conquered over the Platonic organic metaphor, and with it goes talk of final causes and ends and such things. Machines considered as things we use have purposes. The Newcomen engine is for pumping water, which we want it to do. Machines considered as machines—as in science—are just matter in motion, governed by eternal, unchanging laws, and have no purpose. Therefore, there is no purpose to the Newcomen-like cycling globe. It just happens. Lyell, in particular, was ever keen to deny biological progress. He certainly thought that the creation of humans was something special and did demand a divine intervention—a position Darwin erased—but he did not want to see us as the apotheosis of a supposedly blind process. The teleology of the organic model or metaphor is gone forever.

FIGURE 7.1 Newcomen engine. *Credit:* Reproduced from Ephraim Chambers, with supplement by Abraham Rees, *Cyclopedia* (London: J. F. & C. Rivington, 1786).

AGRICULTURE

Organisms live on the surface of a machine. Thanks particularly to our organicists, we know that as the surface shapes us, so we shape the surface—niche construction! Without getting into discussion about who has had the biggest influence, clearly humans are one species that has massively changed our world, and at the same time our world has posed massive challenges to us (Fry 2013). It all began with agriculture, or let us say more cautiously, it was the coming of agriculture that started us toward the modern world with its opportunities and difficulties. We can date the start to around 12,000 years ago, when a massive ice age made the hunter-gathering lifestyle much more difficult, and there was a move to cultivation of crops and the tending of animals—cattle, sheep, pigs, chickens, rice, wheat, potatoes, and more. Tools and simple machines were devised, to grind corn for example, and the result was explosive. No secret—"farmers pump out babies much faster than hunter-gatherers" (Lieberman 2013, 188). Paradoxically, far from large families being drains on the resources, they are the very opposite. "After a few years of care, a farmer's children can work in the fields and in the home, helping to take care of crops, herd animals, mind younger children, and process food. In fact, a large part of the success of farming is that farmers breed their own labor force more effectively than hunter-gatherers which pumps energy back into the system, driving up fertility rates" (188).

Numbers increased exponentially. The human population count went from around five or six million when agriculture started until, 10,000 years later at the time of the birth of Jesus, there were 600 million of us. And that was just a prologue. It took a mere 2,000 years or less to double that number, and then things really got underway. There are around seven or eight billion humans alive today, and we are adding about eighty-three million more every year—with consequences that bring us right up against the issues that opened this chapter. Rather than attempt a general survey—fated to be inadequate—we shall highlight two controversial issues where mechanists and organicists differ, often very bitterly: the Green Revolution and global warming. The former is an enterprise aimed at improving things with success claimed by the mechanists and failure by the organicists, and global warming is something that organicists claim is a bad thing brought on by the mechanists, whereas the mechanists argue that although global warming is undoubtedly bad, the way forward is by more mechanism rather than less.

THE GREEN REVOLUTION

The thinking and the work of the Green Revolution was as firmly located within the machine paradigm as it could possibly have been. It was a movement from around 1950 to around 1970, but with ongoing issues and controversies, aimed at improving agricultural yields dramatically, to speak to widespread poverty and hunger, particularly in developing countries (Jain 2010). There were several related strategies. These included the use of chemical fertilizers, agrochemicals (above all, pesticides), far more sophisticated methods of controlled water-supply (usually involving irrigation), and, most particularly, mechanization of methods of cultivation (using machines for reaping and the like). Above all—at least that which attracted most comment—was the creation of high-yielding varieties (HYVs) of cereals, especially dwarf wheat and rice. In the late 1960s, the Philippines, for example, introduced a new rice cultivar (plant variety produced by selective breeding). Obtained by crossing already-existing varieties, IR8 was a semi-dwarf variety—necessarily (artificially) dwarfed to avoid falling over from increased absorption of nitrogen. It demanded the heavy use of fertilizers and pesticides. In return, one got far higher yields from IR8 than from traditional cultivars. The results really were quite staggering: in two decades, rice production in the Philippines increased from 3.7 to 7.7 million tons per year. Thanks to IR8, for the first time in the twentieth century, the Philippines became a rice exporter. India also felt the benefits of adopting IR8. Using this new strain, you can get about five tons of rice per hectare without fertilizer and about twice that with fertilizer. This is about ten times the amount of normal rice grown under traditional conditions. Like the Philippines, India is now a rice exporter. Also worth mentioning is that the price of rice for domestic consumption has fallen to less than half what it used to cost.

As agriculturalists got better and better insights into the potentials of the molecular understanding of organisms, so ever more sophisticated techniques were used to produce new lines and strains. Thus came about the possibility of the "genetic modification" of organisms (GM foods) as genes were artificially added or removed from natural varieties. Famous, notorious, is Golden Rice (Regis 2019). This is a variety of rice (*Oryza sativa*) modified to produce, through genetic engineering, in the edible parts of rice, more beta-carotene (a strongly colored red-orange pigment abundant in fungi). The great significance is that this is a precursor of vitamin A. Given that rice is the staple crop for more than half of the world's population, and given that this new kind of rice can prevent vitamin A deficiency, its potential for good needs no stress—not in a world where the lack of vitamin A is thought to cause nearly 700,000 early childhood deaths and another 500,000 cases of irreversible blindness.

The story of Golden Rice does not stop there. Expectedly, more work has been done on the rice, and expectedly it is already much improved. Golden Rice 2, announced in 2005, can produce up to twenty-three times as much beta-carotene as the original modified rice. Little wonder that it has been approved for growth and consumption in New Zealand, Australia, Canada, and the United States. And yet, the hostility to Golden Rice in many developing countries is simply staggering! India is but one of many countries that refuse to allow the growth and use of Golden Rice. Why the opposition? When the Green Revolution started, there were understandable concerns about such issues as pollution. You spray everything with pesticides, there are bound to be some ill effects—sometimes major ill effects. Most obviously, you are going to threaten people's health and destroy other organisms that are an important part of daily diets:

> The increasing production of staples displaced the raising of local fruits, vegetables and legumes that are major sources of micronutrients. Due to the loss of diversified crops in the farmers' fields, micronutrient (such as iron, zinc, vitamin A, selenium, iodine, etc.) deficiencies had become notable.
>
> High-yielding Varieties are more prone to pests and diseases compared with traditional cultivars, thus, requiring high level of pesticides. The incessant planting of a few genetically related and similar HYVs, often under double or triple cropping over a wide area, had led to the appearance of new biotypes of insect pests.
>
> Heavy fertilizer applications are producing nitrate levels in drinking water that exceed tolerable levels, while pesticides have eliminated fish and weedy green vegetables from the fields and thus in the diet of poor farmers.

And there are knock-on effects:

> Mechanical pumps, tractors, threshers, reapers and other machineries contribute immensely to raising yield and output, but there is considerable evidence that their net effect in employment is labor displacing. It also removed an important source of employment from the rural economy,

> hence, pushing further down rural wages and encouraging exploitative labor practices. (Pilipinas et al. 2007)

When we come to genetic modification, the criticisms intensify. Indian activist Vandana Shiva (2001) writes of Golden Rice that its use "is a recipe for creating hunger and malnutrition, not solving it." Of the claim that it can reduce significantly problems of blindness affecting children in developing countries, she writes "it appears as if the world's top scientists suffer a more severe form of blindness than children in poor countries." The statement that "traditional breeding has been unsuccessful in producing crops high in vitamin A" is not true given "the diversity of plants and crops that Third World farmers, especially women have bred and used which are rich sources of vitamin A such as coriander, amaranth, carrot, pumpkin, mango, jackfruit." Note that driving Shiva as much as the technological questions is her philosophy—that we need to return to traditional methods that are more in tune with custom and nature (Specter 2014). Explicitly, she acknowledges that she approaches these questions from a holistic perspective. She also links this to "ecofeminism"—a topic to which we shall turn shortly.

Holism triumphant.

GLOBAL WARMING

Our aim above has not been to give a full account of the Green Revolution, its enthusiasts, and its critics. It is enough to show that there are strong divides, and some disputes—many—are driven by philosophical differences as much as hard facts. The same is true also of global warming (Weart 2008). Although the topic of global warming is something that generates much controversy, with opposition particularly from those with a commercial stake in denying it—hydrocarbon energy companies most obviously—and from right-wing politicians appealing to their constituency of anti-intellectual populists, nigh unanimous scientific opinion is that the globe is warming due to human activity, and that this is already proving detrimental to our well-being and has prospects of being absolutely crushing if attention is not paid to it.

Non-controversially, the facts are these. First, an increase in global temperature. There have always been fluctuations in the general temperature of the globe, or at least parts of the globe. Ice ages are well documented. It seems now we are in the reverse of an ice age. The last 150 years have seen a significant rise in the global temperature. At the same time, the sea level has risen, and the snow cover of the Northern Hemisphere—especially up close to the Arctic—has decreased. Reasonable estimates are that between 1880 and 2012, the surface temperature of the Earth has increased by about 1°C, and rather more if you put the starting date back to the eighteenth century and the beginning of the Industrial Revolution.

What is striking is that this rise in temperature is not natural in the sense of being caused by nonhuman forces. For instance, it is suspected that natural climate changes might follow on the shifting of the land–sea balance thanks to plate tectonics. In the case of the recent rise, that which we are now experiencing, almost all agree that it is brought on by human activities, and that if we continue on course, we can expect a speeding up of temperature rise, perhaps by 2°C or more by the end of this century, 2100. There is debate about the exact causes and their relative importance, but general agreement is that three factors are very significant (Mathez & Smerdon 2018). The first is the ever-greater reliance on fossil fuels. In order for it to be habitable at all, the Earth needs its atmosphere containing the "greenhouse gases," which prevent the heat from the Sun dissipating almost immediately. As in a greenhouse, solar heat is trapped, and the general temperature of the Earth rises, enough to bear life. The chief greenhouse gas is carbon dioxide, produced naturally by animals and by geological phenomena such as volcanoes. The use of fossil fuels produces carbon dioxide artificially, and it too enters the atmosphere and traps energy from the Sun, with consequent temperature rise. Second, as is well known, plants absorb carbon dioxide. That is why, naturally, there is a balance. However, human activity disrupts this and destroys the balance. Most obviously there is the destruction of forests and jungles as in Brazil. More carbon dioxide remains in the atmosphere. Third is the use of aerosols. Their emissions get into the atmosphere, and some of them too add to the increased trapping of the energy from sunlight.

It is nigh tautological that global warming has elicited responses that again show the divide between mechanistic and organismic thinking. On the one side, the immediate response is not to give up and try to return to the eighteenth century, but rather to see how mechanistic science can alleviate the problems. One obvious way is to turn to alternative sources of energy, reducing our reliance on fossil fuels. This is already happening, if in a limited way. For instance, in the United States and a number of other countries, the use of coal is declining dramatically. True, much of this is due to turns to other fossil fuels, natural gas for instance. More controversial has been the use of nuclear methods of producing electricity. Less tense-making, wind and solar power are starting to make a real difference; in the UK in 2020, about 40 percent of the electricity produced came from renewable production. Florida prides itself on being the "sunshine state." Finally, it is being realized that this might point to a resource, and now around the state, solar farms are appearing at a steady rate. At the same time, shipments of coal from mining states have dropped dramatically. In the twenty years that he has lived in Florida, one of the authors has moved entirely from relying on coal-produced electricity to solar-produced energy.

On the other side, as expectedly, there have been calls for holistic approaches. Not everyone suggests that we all move to North Dakota, live in yurts, ride bicycles, eat vegan, and—in order to control the population explosion—practice only eyesight-threatening sexual activity, although there is often a flavor of this. More, it is an exhortation to think in an organic sort of way, where everything is integrated into a joint interdisciplinary approach. Pope Francis is one of the most urgent and articulate of this kind of thinking. In 2015, he issued an encyclical letter on caring for the environment, condemning what we humans have done and pleading for a change:

> 1 "Laudato si', mi' Signore"—"Praise be to you, my Lord." In the words of this beautiful canticle, Saint Francis of Assisi reminds us that our common home is like a sister with whom we share our life and a beautiful mother who opens her arms to embrace us. "Praise be to you, my Lord, through our Sister, Mother Earth, who sustains and governs us, and who produces various fruit with coloured flowers and herbs."
>
> 2. This sister now cries out to us because of the harm we have inflicted on her by our irresponsible use and abuse of the goods

> with which God has endowed her. We have come to see ourselves as her lords and masters, entitled to plunder her at will. The violence present in our hearts, wounded by sin, is also reflected in the symptoms of sickness evident in the soil, in the water, in the air and in all forms of life. This is why the earth herself, burdened and laid waste, is among the most abandoned and maltreated of our poor; she "groans in travail" (Rom 8:22). We have forgotten that we ourselves are dust of the earth (cf. Gen 2:7); our very bodies are made up of her elements, we breathe her air and we receive life and refreshment from her waters. (Francis 2015)

Enough said!

ENVIRONMENTALISM

We are already moving from presenting mechanism alone to presenting organicism as an alternative. Hence, as is our wont, let us pull back now, and as we have considered the machine metaphor approach from a historical perspective, let us do the same for the organic metaphor approach. In many respects, more than for mechanism, organicism is tailor-made for thinking about environmental issues:

> Our environmental ethic will be more holistic. Pains and pleasures will be part of a larger picture, derivative from and instrumental to further values at the ecosystemic level, where nature evolves a flourishing community in some indifference to the pains and pleasures of individuals, even though pain and pleasure in the higher forms is a major evolutionary achievement. (Rolston, 1988, 108)

The very crux of the metaphor is that the Earth is an organism. Goethe's thinking meshes nicely:

> Whatever Nature undertakes, she can only accomplish it in a sequence. She never makes a leap. For example she could not produce a horse if it were

> not preceded by all the other animals on which she ascends to the horse's structure as if on the rungs of a ladder. Thus every one thing exists for the sake of all things and all for the sake of one; for the one is of course the all as well. Nature, despite her seeming diversity, is always a unity, a whole; and thus, when she manifests herself in any part of that whole, the rest must serve as a basis for that particular manifestation, and the latter must have a relationship to the rest of the system. (Naydler 1996, 60)

As always with the organic metaphor, you find value rather than confer it. "The land ethic rests upon the discovery of certain values—integrity, projective creativity, life support, community—already present in ecosystems, and it imposes an obligation to act so as to maintain these" (Rolston 1988, 228).

Skipping past pre–Scientific Revolution thinking, from the beginning of the Romantic movement, we find that environmental issues have indeed been front and center for organicists. Nowhere did this hold more truly than in the adolescent United States. Remember the New England "Transcendentalists," into which the newly arrived Louis Agassiz fit so comfortably. Listen to Henry David Thoreau, best known for his book *Walden, or Life in the Woods* (1854), where he tells of two years of simple living in the 1840s in a hut he had built on the shores of Waldon Pond, on land owned by his friend and fellow transcendentalist, Ralph Waldo Emerson:

> There is nothing inorganic. These foliaceous heaps lie along the bank like the slag of a furnace, showing that Nature is "in full blast" within. The earth is not a mere fragment of dead history, stratum upon stratum like the leaves of a book, to be studied by geologists and antiquaries chiefly, but living poetry like the leaves of a tree, which precede flowers and fruit—not a fossil earth, but a living earth; compared with whose great central life all animal and vegetable life is merely parasitic. Its throes will heave our exuviae from their graves. ("Spring," Thoreau 1854, 2, 476)

A movement was started. A few years later, we have the philosophy of Scottish-born John Muir, the founder of the Sierra Club. He was a kind of pantheist, following the Dutch, seventeenth-century philosopher

Baruch Spinoza in identifying God with the creation. "Beauty is God, and what shall we say of God that we may not say of Beauty?" (Notebook 1872, Worster 2008, 208). Although Muir became a Darwinian, he preferred a softer, holistic version. "I never saw one drop of blood, one red stain on all this wilderness. Even death is in harmony here" (Muir 1966, 93). In the first part of the twentieth century, the hero of the environmentalist movement was the Wisconsin land-and-game manager, Aldo Leopold. In 1923, Leopold wrote that many of us "have felt intuitively that there existed between man and the earth a closer and deeper relation than would necessarily follow the mechanistic conception of the earth as our physical provider and abiding place" (Leopold 1979, 139). Adding: "Philosophy, then, suggests one reason why we can not destroy the earth with moral impunity; namely, that the 'dead' earth is an organism possessing a certain kind and degree of life, which we intuitively respect as such" (140). It was this philosophy that grounded his very popular *A Sand County Almanac,* published (in 1949) posthumously after he died fighting a forest fire.

Expectedly, Alfred North Whitehead and his followers were both cause and effect of this movement. The organismic view of nature was the very essence of Whitehead's philosophy. In *Science and the Modern World,* he wrote:

> The trees in a Brazilian forest depend upon the association of various species of organisms, each of which is mutually dependent on the other species. A single tree by itself is dependent upon all the adverse chances of shifting circumstances. The wind stunts it: the variations in temperature check its foliage: the rains denude its soil: its leaves are blown away and are lost for the purpose of fertilisation. You may obtain individual specimens of fine trees either in exceptional circumstances, or where human cultivation has intervened. But in nature the normal way in which trees flourish is by their association in a forest. Each tree may lose something of its individual perfection of growth, but they mutually assist each other in preserving the conditions for survival. The soil is preserved and shaded; and the microbes necessary for its fertility are neither scorched, nor frozen, nor washed away. A forest is the triumph of the organisation of mutually dependent species. (Whitehead 1926, 206)

SILENT SPRING

Both reinforcing the continued enthusiasm for the organic model of the world and stressing its value content—organicism is morally better than mechanism—we have one of the more curious and little-known stories in this whole book. The greatest environmentalist book of the mid-twentieth century—the greatest environmentalist book of the whole century—was Rachel Carson's *Silent Spring,* first published in 1962. A skilled science writer, Carson had long worked within a Whiteheadian organismic world-view. Typical is a passage from an earlier book, *The Sea Around Us:*

> So too, the lifelessness, the hopelessness, the despair of the winter sea are an illusion. Everywhere are the assurances that the cycle has come to the full, containing the means of its own renewal. There is the promise of a new spring in the very iciness of the winter sea, in the chilling of the water, which must, before many weeks, become so heavy that it will plunge downward, precipitating the overturn that is the first act in the drama of spring. (Carson [1950] 2018, 35–36)

Now, in *Silent Spring,* Carson brought her formidable forensic abilities to bear on the havoc being created in the environment by man-made substances—chemicals! She seized on and detailed the devastating waste being laid by dichlorodiphenyltrichloroethane (DDT)—for all that, it was apparently a godsend, given the way it combated mosquitoes, the spreaders of malaria. Again and again, Carson's language suggests strongly that she is working from within the organic model (Norwood 1987). She gave one chapter the heading: "Earth's green mantle" and another "Nature fights back." Her writing is drenched in values writing of "the obligation to endure" (Carson 1962, 13). Overall, she is working with the analogy between the way we are wrecking our planetary home and the way we are wrecking our own bodies. Later, for her funeral service, she chose readings of the same ilk and tenor, telling us that as we study nature "we come to perceive life as a force as tangible as any of the physical realities of the sea, a force strong and purposeful, as incapable of being crushed or diverted from its ends as the rising tide" (Carson 1965, 250).

We said that Carson was a skilled science writer. She knew full well that her book was going to arouse the ire of the chemical industry, which was making very good money out of the production of substances such as DDT. They were going to push back, and one effective way would be to portray Carson as a flaky environmentalist—a do-gooder out of touch with reality. The fact that she was a woman meant that already half the job was done for them. Completely sensitive to all this, Carson wrote *Silent Spring* in a very controlled manner, being very careful not to give out grist for the mill (Ruse 2013). Although she was an organicist-holist, she never said that explicitly in her book. Critics would have jumped all over such foggy thinking. She was equally careful about revealing her sources. Carson was one of an informal group of mutually supportive, activist women, one of whom was Marjorie Spock, incidentally the younger sister of Benjamin Spock, the famous pediatrician. Marjorie Spock was an enthusiastic anthroposophist, a follower of a man introduced in Chapter 4, the Austrian clairvoyant and polymath Rudolf Steiner. Best known as the founder of the Waldorf system of education, Steiner dabbled in many other things, including architecture, painting, writing, economics, medicine, and—relevant here—gardening and agriculture. He was the creator of so-called biodynamic farming, a kind of organic farming with esoteric trimmings, such as the need to use fertilizers made from cow horns filled with manure planted in the ground to mature over winter. The horns of bulls spoil the whole process. Steiner was much influenced by Goethe and by Schelling also somewhat, and this naturally reflects into a full-blooded organicist view of the world, albeit somewhat idiosyncratic in that apparently Earth's organs appear on the outside of the skin rather than inside. "If from this point of view we now compare the Earth's surface with the human diaphragm, then we must say: In the individuality with which we are here concerned, the head is *beneath* the surface of the Earth, while we, with all the animals, are living in the creature's belly! Whatever is *above* the Earth, belongs in truth to the intestines of the 'agricultural individuality,' if we may coin the phrase" (Steiner 1924, 30).

Steiner's followers, to a person, accepted all of this as gospel, as one might say. One enthusiast, a leading bio-dynamic agriculturalist in the United States, Ehrenfried E. Pfeiffer, took up the pesticide issue with vigor in the late 1950s, publishing (in the house journal) an article with the revealing title "Do we know what we are doing? DDT spray programs—their values and dangers" (Pfeiffer 1958). Within a month, courtesy of Marjorie Spock, it was in Carson's hands, who replied with thanks noting that it was a "gold mine." It certainly was. All of Pfeiffer's arguments appeared virtually unchanged in *Silent Spring*—appeared unacknowledged, also. Revealing the source would have been suicide.

THE GAIA HYPOTHESIS

The moral of this little-known story of the backdrop to Carson's views is that the organicist world picture had a huge, if often unnoted, influence in the twentieth century. And it was bound up all the way with value. Mechanical thinking bad; organic thinking good. This does not mean that the picture was uncontroversial or that more conventional scientists—more conventional, post-Darwinian biologists—did not fight back. This is all well illustrated by the controversy over the Gaia hypothesis that broke out in the 1970s. For all that James (Jim) Lovelock trained as an organic chemist, he showed his true brilliance in the 1940s and 1950s by his genius as an inventor (Lovelock 2000). He could take a pile of cast-off electronic junk—often army surplus—and create the most powerful and significant machines. One of his great triumphs was to create an instrument, the electron capture detector (ECD), so precise that, to use an example of Lovelock himself, it can record in Britain within a week or two the effects of a bottle of solvent spilt on a cloth in Japan. More practically and pertinently, it led to the discovery of the widespread existence of chlorofluorocarbons (CFCs) in the atmosphere. Among other things, this kind of achievement led to Lovelock's fellowship of the Royal Society (of London). It also led to general panic about how CFCs—from certain aerosols—were contributing to the shrinking ozone layer above the Earth.

In the early 1960s, Lovelock was working for NASA in California on the problem of detecting life on Mars. Always a maverick, to the disappointment of the many who liked that sort of thing, he argued that there is no need to build expensive spaceships to go to Mars and study the planet at first hand. By studying the atmosphere, you can do nearly all you need to do down here on Earth. If there are no chemicals such as oxygen and methane, you are certainly not going to get life on Mars:

> As Pasteur and others have said, "Chance favours the prepared mind." My mind was well prepared emotionally and scientifically and it dawned on me that somehow life was regulating climate as well as chemistry. Suddenly the image of the Earth as a living organism able to regulate its temperature and chemistry at a comfortable steady state emerged in my mind. At such moments, there is no time or place for such niceties as the qualification "of course it is not alive—it merely behaves as if it were." (Lovelock 2000, 253–254)

Jim Lovelock is a polished raconteur, and he always tells a good story. Whatever the fine details, something did happen, and he did become convinced of the organic nature of our Earth—note, although Platonic, in spirit a little different from the vision of the Greek philosopher. For Plato, the organism is the whole universe. For Lovelock, the very point is that it is the Earth alone that is an organism. Lovelock makes much of the somewhat culture-deprived background from which he emerged. Despite the romancing to which this leads, it is probably true that there was no direct Platonic influence. What is almost certainly true is the direct influence of—once again—our friend Rudolf Steiner! Somewhat of a loner, in the 1960s, Lovelock was living in a rural English village. His only constant intellectual contact (and friend) was his fellow pub mate, none other than the future winner of the Nobel Prize in Literature, William Golding, then hugely popular as the author of *The Lord of the Flies*. Golding had for many years been an enthusiastic anthroposophist and a follower of Steiner. Although by then he was no longer quite so committed, he still accepted much of the thinking, including the hylozoism. This came through a few years later in his Nobel acceptance speech. We "have been caught up to see our earth, our mother, Gaia Mater, set like a jewel in space. We have no excuse now for supposing her riches inexhaustible nor the area we have to live on limitless because unbounded. We are the children of that great blue white jewel. Through our mother we are part of the solar system and part through that of the whole universe. In the blazing poetry of the fact we are children of the stars" (Golding 1983).

As you might expect, later, Lovelock was not particularly keen to acknowledge any debt to Steiner. He had been battered enough as it is. But we know that Gaia was much talked of by Lovelock and Golding. "When I first discussed it with Bill Golding, we went into it in considerable depth" (interview of Jim Lovelock by Michael Ruse, January 18, 2011). Indeed, it was Golding that came up with the name "Gaia." We have just seen that Golding carried on the Steiner thinking about the world. And we know that Steiner was discussed by the two men favorably. On Golding's recommendation, Lovelock sent one of his kids to a Steiner (Waldorf) school and was much satisfied. It all fits, particularly that Lovelock was entering unknown and dangerous territory. To have some real backing from someone to be respected, even if unacknowledged, was a major something.

In the early 1970s, Lovelock was ready to go public (Lovelock & Margulis 1974a, 1974b). He had acquired a co-author, Lynn Margulis, just then rising to fame as the promoter of the hybrid origin of the eukaryotic cell. She argued that it came from the fusion of simpler prokaryotic cells. Highly controversial at the time, she was later vindicated by molecular studies, and rightfully she was inducted into the National Academy of Sciences—an honor never awarded to her first husband, the science popularizer Carl Sagan. However, although always supportive, Margulis was never the main player, and she soon went off following follies of her own.

Staying on track, what is to be said of pertinence to us? First, Gaia was not stupid. Lovelock and Margulis may have been mavericks. They were good scientists. No one questioned the propriety of their being in their countries' leading science organizations. There was some evidence for Gaia, although here, as often, it depended on what you meant by "evidence." It is true that there is a train that runs from London to Paris, but it is not exactly what you would count as evidence for Gaia. Lovelock was by training a chemist, and although he had much experience working on issues involving humans—for example the spread of germs among bomber pilots in the Second World War—he thought as a chemist. For him, what really counted for something to be an organism was homeostasis—a body being able to keep itself in balance. Not a silly thought when you consider. Humans keep a constant body temperature through sweating and shivering. A lump of rock does not. He and Margolis pointed out that, considering heat from the Sun and the cooling of the Earth through heat being lost, we should first have a cooling of the Earth since its formation and then a rise. What we have in fact is an almost stable temperature, suggesting that somehow the Earth is in balance. Homeostasis! What better proof could one have that the Earth is an organism?

Second, in the opinion of most scientists, Lovelock thought and wrote in ways that were at best old-fashioned or ignorant and at worst perverse. There was often a teleological flavor to his thinking that would have upset Robert Boyle—perhaps not Kepler!—let alone today's scientists. Somewhat disappointed that the scientific community did not take up his hypothesis with enthusiasm, Lovelock turned to speak to the general public. In his *Gaia: A New Look at Life on Earth,* published in 1979, he felt able openly to talk in terms of purposes and ends. That is, after all, the defining feature of organisms. He wrote that "if Gaia does exist, then we may find ourselves and all other living beings to be parts and partners of a vast being who in her entirety has the power to maintain our planet as a fit and comfortable habitat for life" (1). Marshaling the evidence, one powerful item is the fact that the salinity of the sea has apparently been stable, whereas one might have expected it to get more and more salty, like the Dead Sea. To explain this, Lovelock referred to normal, mechanistic processes, fueled by efficient causes, working in a kind of feedback mode. Linking the salt level of the sea to the amount of silica in the water, Lovelock wrote of one such process for keeping things stable:

> This biological process for the use and disposal of silica can be seen as an efficient mechanism for controlling its level in the sea. If, for example, increasing amounts of silica were being washed into the sea from the rivers, the diatom population would expand (provided that sufficient nitrate and sulfate nutrients were also in good supply) and reduce the dissolved silica level. If this level fell below normal requirements, the diatom population would contract until the silica content of the surface waters had built up again, and this is well known to occur. (88–90)

All very conventional. But then, at once, Lovelock showed that he, *as a scientist,* was putting this in a teleological context. "This deluge of dead organisms [diatoms] is not so much a funeral procession as a conveyor belt constructed by Gaia to convey parts from the construction zone at surface levels to the storage regions below the seas and continents." To spell it out: "CONSTRUCTED by Gaia IN ORDER TO convey parts."

FELLOW TRAVELERS

Third, the general public loved this kind of stuff (Ruse 2013). It took a while for things to catch fire, but by the end of the 1970s, Gaia was—and remains—a much-discussed and embraced idea. Google "Gaia" and see how many spin offs there are, especially in California: Gaia herbs (you can get these at Walmart!), Gaia yoga ("explore transformation through consciousness expansion with the luminaries on Gaia"), Gaia Fairtrade ("an organization founded by women to empower women"). Let us be fair. The East Coast enthused no less than the West Coast. Appropriately, as we move from the land of the hippies to the land of the stuffed shirts, we find that for some reason, Gaia seems to catch the fancy of Episcopalians (Anglicans). Notable was a celebration of Mother's Day in 1981 in the New York City Cathedral of Saint John the Divine. It featured *Missa Gaia,* an ecological and ecumenical mass, with music composed by composer Paul Winter. "The Alaskan tundra wolf whose voice this Kyrie was based on, sings the same four-note howl seven times in an interval known as the tritone—the sax, tenor solo voices and chorus answering" (program notes).

Anyone who doubts we are dealing with value issues should start here—a point that keys us to the fact that by the 1970s there were other related movements pushing the organicist environmental thesis. Prominent were the so-called deep ecologists. Led by the Norwegian philosopher, Arne Næss—he acknowledged the strong influence of *Silent Spring*—he trumpeted the importance of the organicist perspective: "The well being and flourishing of human and non-human life on Earth have value in themselves (synonyms: intrinsic value, inherent worth). These values are independent of the usefulness of the non-human world—for human purposes" (Næss 1986a, 68). Life here means more than just organisms, the biosphere. It is the whole world *around* us. "The term 'life' is used here in a more comprehensive non-technical way also to refer to what biologists classify as 'non-living': rivers (watersheds), landscapes, mountains, oceans. For supporters of deep ecology, slogans such as 'let the river live' illustrate this broader usage so common in many cultures" (68).

As is the wont of thinkers such as this, the mechanical approach is anathematized. Speaking of mechanism as "the view of the universe as a mechanical system composed of elementary building blocks, the view of the human body as a machine, the view of life in society as a competitive struggle for existence, the belief in unlimited material progress to be achieved through economic and technological growth, and last but not least, the belief that a society in which the female is everywhere subsumed under the male is one that follows a basic law of nature," it is sadly concluded that it will not do. We need a "fundamental change of world view" (Capra 1987, 19–20). "Competitive struggle," "unlimited material progress," "female everywhere subsumed under the male." Need we say more?! The world is a harmonious entity with its own intrinsic worth. "I suspect that our thinking need not proceed from the notion of living being to that of the world, but we will conceive reality, or the world we live in, as alive in a wide, not easily defined, sense. There will then be no non-living beings to care for" (Næss 1986b, 234).

"Subsumed under the male"! Little wonder that alongside and overlapping deep ecology, was another movement: ecofeminism. The idea of the eternal feminine—giving and caring—is nigh commonsensical. Mother Earth!

> She sweeps with many-colored Brooms—
> And leaves the Shreds behind—
> Oh Housewife in the Evening West—
> Come back, and dust the Pond!
> (Emily Dickinson, Poem 219, 1891)

With the 1960s rise of feminism generally, it was natural that attention should turn to environmental issues. Setting the background, as it were, the historian Caroline Merchant has been very influential; in her deservedly praised book, *The Death of Nature,* she sees, in the move from organicism to mechanism, the perhaps-unintended consequence of downgrading the nature and role of women. We start with nurturing Mother Earth, and then mechanism kills her. "Between the sixteenth and seventeenth centuries the image of an organic cosmos with a living female earth at its center gave way to a mechanistic world view in which nature was reconstructed as dead and passive, to be dominated and controlled by humans" (Merchant 1980, xvi). This at once reflects on the status of women. "If women overtly identify with nature and both are devalued in modern Western culture, don't such efforts work against women's prospects for their own liberation?" (xvi). Expectedly, Darwinism has a role in this sad story. "In the nineteenth century, Darwinian theory was found to hold social implications for women. Variability, the basis for evolutionary progress, was correlated with a greater spread of physical and mental variation in males. Scientists compared male and female cranial sizes and brain parts in the effort to demonstrate the existence of sexual differences that would explain female intellectual inferiority and emotional temperament" (162–163). The hope is that present-day science can be refashioned along more female-friendly lines. "Ecology, as a philosophy of nature, has roots in organicism—the idea that the cosmos is an organic entity, growing and developing from within, in an integrated unity of structure and function" (100).

Fast forwarding to the present, predictably we find that it is men to blame for everything. "The physical rape of women by men in this culture is easily paralleled by our rapacious attitudes toward the Earth itself. She, too, is female. With no sense of consequence in the scant knowledge of harmony, we gluttonously consume and misdirect scarce planetary resources" (Razak 1990, 165). As predictable is the linking of sexuality with organicism. "The experiences inherent in women's sexuality are expressions of the essential, holistic nature of life on Earth; they are 'body parables' of the profound oneness and interconnectedness of all matter / energy, which physicists have discovered in recent decades at the subatomic level" (Spretnak 1989, 129). The phallocentric nature of so much environmental thinking is a constant theme. Thanks to Darwin, "the symbioses, the interconnections that nurture and sustain life are ignored, and both natural evolution and social dynamics are perceived as impelled by a constant struggle of the stronger against the weaker, of constant warfare." We need an ecofeminist perspective. "Only in this way can we be enabled to preserve and respect the diversity of all life forms, including their cultural expressions, as true sources of our wellbeing and happiness" (Miles & Shiva 2014, 6).

At such a point, it seems almost mean to point out that much of the underlying philosophy comes straight from Plato:

> From one perspective, we realise that we need food, shelter, and clothing; from another that some sort of relationship among people, animals, and the Earth is necessary; from another that we must determine our identity as creatures not only of our immediate habitat but of the world and the universe; from another that the subtle, suprarational reaches of mind can reveal the true nature of being: All is One, all forms of existence are comprised of one continuous dance of matter / energy arising and falling away, arising and falling away. Only the illusions of separation divide us. The experience of union with the One has been called cosmic consciousness, God consciousness, knowing the One Mind, etc." (Razak 1990, 127)

It is true that (in *The Republic*), Plato was prepared to allow for female philosopher kings in his ideal state, but let there be no doubt as to the proper ordering: "Women and men have the same nature in respect to the guardianship of the state, save insofar as the one is weaker and the other is stronger."

It is tempting to let this discussion get side-tracked into listing movements in the 1970s that meshed with Gaia and often drew support from Lovelock's speculations. Unsurprisingly, out in California the pagans—or neo-pagans as they prefer to be called—loved this kind of stuff. When they are not active publicly or privately "sky clad" (stark naked), they turn to the MEANING OF IT ALL. Thus, Oberon Zell-Ravenheart, "initiate in the Egyptian Church of the Eternal Source" as well as "a Priest in the Fellowship of Isis," using the term "Terrabios," which he later identified with Gaia: "it is a biological fact (not a theory, not an opinion) that ALL LIFE ON EARTH COMPRISES ONE SINGLE LIVING ORGANISM! Literally, we are *all* 'One'" (Zell-Ravenheart 2009, 92). Continuing: "The blue whale and the redwood tree are not the largest living organisms on Earth; the ENTIRE PLANETARY BIOSPHERE is." Individual organisms are the cells of Terrabios. The natural habitats (biomes)—deserts and the forests and the prairies and the coral reefs and the like—are the organs. "ALL the components of a biome are essential to its proper functioning, and each biome is essential to the proper functioning of Terrabios."

UNHAPPY MECHANISTS

Just a list—Gaia, deep ecology, ecofeminism, neo-paganism—but the value of such a list is the extent to which it shows that the organicist movement was one that appealed to the general public, and the reasons were as much (if not more) moral than epistemological or scientific. Organicism / holism is in some sense intrinsically good—caring, all-embracing, life sustaining—and mechanism / reductionism is intrinsically bad—indifferent, separating, life destroying, which to a great extent goes far to explain the violent reaction by the professional scientific community to this kind of stuff, especially the violent reaction to Jim Lovelock. As a professional scientist himself, he was thought to be akin to Judas Iscariot, not least for having gone public and thus demanding an explicit response rather than condescending silence. In the opinion of right-minded scientists, the mixing of science and values was anathema. Richard Dawkins was incandescent. He had nothing against interrelatedness as such. In the *Extended Phenotype,* he gave many elaborate examples of natural selection promoting such phenomena. "A beaver that lives by a stream quickly exhausts the supply of food trees living along the stream bank within reasonable distance. By building a dam across the stream the beaver quickly creates a large shoreline which is available for safe and easy foraging without the beaver having to make long and difficult journeys overland" (Dawkins 1982, 200). It was the teleological flavor of Lovelock's thinking that stuck in Dawkins's craw. As a Darwinian, Dawkins fully accepted final-cause thinking. It was just that it had to be a result of natural selection and could not just be the result of some kind of internal force—let alone an external designer:

> The fatal flaw in Lovelock's hypothesis would instantly have occurred to him if he had wondered about the level of natural selection process which would be required in order to produce the earth's supposed adaptations.

> Homeostatic adaptations in individual bodies evolve because individuals with improved homeostatic apparatus pass on their genes more effectively than individuals with inferior homeostatic apparatuses. For the analogy to apply strictly, there would have to have been a set of rival Gaias, presumably on different planets. Biospheres which did not develop efficient homeostatic regulation of their planetary atmospheres tended to go extinct. The Universe would have to be full of dead planets whose homeostatic regulation systems had failed, with, dotted around, a handful of successful, well-regulated planets of which the Earth is one. (235–236)

All of this is quite apart from (the ludicrous idea of) planets reproducing so you can get natural selection to work on them.

It is hard to imagine that anyone could be quite as critical as Richard Dawkins. If there is a Platonic Form of Outraged Mechanist, it is probably he. Others, however, chipped in to do their bit. Canadian biologist W. Ford Doolittle found the group-selection underpinnings of Gaia deeply unsatisfactory. "The construction of an evaporation lagoon for sequestration of sea salt may benefit the biosphere as a whole, in the very long run, but what in particular does it do for the organisms who construct it, especially in the short run?" (Doolittle 1981, 61). This is not to mention dodgy assumptions about the merits of balance. "The global conflagration expected if oxygen levels exceed 25% would be disastrous to most higher forms of life. But it would produce a large amount of carbon dioxide and consume a lot of oxygen, and it is carbon dioxide which is the life-giving substrate for the methanogens and it is oxygen which they must scrupulously avoid (because it is toxic to them). Would methanogens not in fact benefit, at least for thousands of years, from such a disaster?" (61).

Expectedly, Lovelock responded vigorously. He was savvy enough to realize that he would never succeed in a straight defense of organicism. So, he took on the conventional scientists on their own grounds. He gave a machine-metaphor response! Lovelock understood—one doubts he ever really accepted—the complaint about teleology. In the path of Darwin, he set about finding efficient cause adequate to explain the final-cause picture that he had (and has) of planet Earth. One mechanism that Lovelock devised—an early computer model, reflecting his ease with electronic devices—was Daisyworld (Watson & Lovelock 1983). One supposes a planet with two kinds of flowers: white daisies and black daisies. In the sunlight, the white daisies reflect the Sun, and the black daisies absorb the Sun. It turns out that the quicker you heat, the quicker you are to reproduce, so black daisy numbers go up. As this happens, the planet generally heats up, and so the white daisies start to get going sooner, and for various reasons, they outcompete the black daisies. The heating of the globe backs off, and things go on this way until you get a balance. The point—the important point—is that the planet's temperature is balanced—in homeostasis—rather than, as one might expect, constantly heating up. One has the kind of end-directed process that Lovelock sought, without the need of recourse to either a modern-day Demiurge or Aristotelian final causes.

Lovelock has not been alone in this quest. Of all people, William Hamilton—he of kin selection fame—got involved, providing the kind of circular feedback system that Lovelock needed. More and more scientists were now starting to realize that the kinds of circular processes so central to Lovelock's thinking are indeed widespread and in need of explanation. Much effort was put into tracing and explaining the circle that carries around the molecule dimethyl sulfide, something excreted by algae. Hamilton, working with a young researcher, tied all of this in with benefits to the algae, as they help form clouds that bring on winds that disperse the algae (Hamilton & Lenton 2005). Everything is efficient cause. More than this, it is all individual selection. It is the individual algae that benefit, not the species. Hamilton had been very critical of Lovelock on this score, arguing (correctly) that Lovelock was supremely indifferent to the individual / group selection issue, no doubt an attitude born of ignorance. Rather unkindly, Hamilton characterized Lovelock's hypothesis (about the Earth being an organism) as akin to "a grant application in the field of geochemistry that wanted to investigate the quantum status of the earth-moon super-atom system" (letter to Lovelock 1996, quoted in Ruse 2013, 164). Yet, apparently, now from such nonsense thinking emerged a mechanism that did the job and at the same time eschewed both unacceptable teleology and group-selection hypotheses—reductionistic but surely in some very real sense holistic, if not teleological as well.

This kind of work epitomizes the state of things today. There is a thriving professional discipline known as "Earth system science." It looks at interactions between organisms and the environment, focusing heavily on the circular feedback systems we have been discussing. The leading journal *Science,* no less, acknowledged Lovelock's role in getting things moving. "James Lovelock's penetrating insights that a planet with abundant life will have an atmosphere shifted into extreme dynamic disequilibrium, and that Earth is habitable because of complex linkages and feedbacks between the atmosphere, oceans, lands, and biosphere, were major stepping-stones in the emergence of this new science" (Lawton 2001, 1965). This said, there is no place for the kind of teleological thinking that came so naturally to Lovelock. Indeed, even those who are privately supportive of Lovelock's hypothesis are careful to refrain from public talk about the Gaia hypothesis. Mention of it in a prospective journal or conference presentation is reason enough for instant rejection. It is regarded rather like spiritualism or phrenology—how some today regard chiropractic medicine. Not that Lovelock has really changed at all. In a late book, he asked: "Why do we pee? Not so silly a question as it might seem. The need to rid oneself of waste products like salt, urea, creatinine and numerous other scraps of metabolism is obvious but only part of the answer. Perhaps we pee for altruistic reasons. If we and other animals did not pass urine some of the vegetable life of the Earth might be starved of nitrogen" (Lovelock 2006, 19). We leave the full understanding of this passage as an exercise for the reader. It is nice to think that after a night in the pub, drunken philosophy students are behaving in ways that benefit us all.

ENVOI

What should we conclude? Somewhat pessimistically—although perhaps by this point, the reader will be moving to the viewpoint of the authors that pessimism is the last emotion we should be feeling; better comprehendingly, even excitedly—the answer seems to be that there is a mechanism/organicism divide over the proper approach to environmental questions, that it is ongoing. Whether this divide is coming to an end is unclear; there are signs that it may be. Mainstream ecology is now much more likely to take seriously holistic understandings of the natural environment that are typically held by indigenous peoples. The field of integral ecology "honors and integrates multiple approaches to the environment. Sees value in all perspectives" (Esbjorn-Hargens & Zimmerman 2009, 227). Many ecologists working in ecosystem management and conservation studies appreciate that one needs to listen to and work with local people, not just tell them what to do to manage or conserve the environment. Values and politics lie at the heart of ecosystem management and conservation, whether one is talking about the protection of fish stocks, looking after forests, or the establishment of nature reserves. It is now commonplace that human health, mental and physical, is helped by our spending time in nature (R. Duncan 2018). At the same time, and returning to the linking theme of this book, the divide is as much if not more philosophical as scientific. However, now we have a new player coming on field and joining the game. Go back to the discussion of science. We have seen that niche construction was an important topic for both mechanists and organicists. Really, however, it is a topic for professionals in the field. The general public knows about beavers and that sort of thing, but the technicalities of their understanding are not that gripping or important.

Environmental issues are different. For the man or woman in the street, they are both gripping and important. Genetically modified foods, global warming, nuclear power to drive power plants—you don't get more relevant. What you might call the sociology of science is up front now. And this is a factor—almost all would agree the most important factor—that often sustains the divide between mechanists and organicists. Recognizing that there will be exceptions, mechanists tend to be practicing professional scientists, schooled in the power of the machine metaphor and finding this confirmed in their daily activities. Organicists, to the contrary, tend to be members of the general public who are concerned about issues directly impinging on their lives, those of themselves and those of others. The power of the machine metaphor means nothing to them, at least at a direct level.

This is only part of the explanation. Why doesn't the general public take the word of professional scientists about the power of the machine metaphor and leave things at that? Simply because in itself the organic metaphor is taken—as we have seen—to be intrinsically better, more morally sensitive, more in tune with "nature" than the machine metaphor. Away with pure reductionism, reducing things including organisms to their parts, indifferent to whether they are alive or dead. In with holism, caring about the individual and the world within which we live. Selfish genes versus Mother Earth.

We are certainly not going to settle the issue here, at the end of this chapter, but let us leave the reader with two reflections. Positively, mechanists would defend strongly their machine-metaphor approach and decry the organic metaphor approach. The Green Revolution controversy shows this full well. Having sex with small children is grossly immoral, unnatural. Why should making small children eat GM foods be any different? Simply, respond the mechanists, because having sex with small children creates incalculable harm—physically and psychologically. Giving small children Golden Rice not only does no harm but can prevent great harm. "These include the burden of night blindness, corneal scars, blindness caused by corneal scars, measles and mortality of children five years old and younger, and night blindness for pregnant and lactating women." Even worse, the cost of failure to make use of Golden Rice "amounts to about 71,600 child deaths annually" (Wesseler & Zilberman 2014, 11). Let us repeat this. Because India has turned its back on Golden Rice, the estimate is that 71,600 children *a year* have died from malnutrition. We condemn Heinrich Himmler as a mass murderer. Why should we do otherwise for Vandana Shiva, a leader among those who condemn Golden Rice?

However, there is also a negative point to be made. Holism is taken as intrinsically good. Ask, first, who were the biggest holists of the twentieth century? The National Socialists! Their whole philosophy was the state as an organic entity with outsiders threatening it:

> All in all, the National Socialistic conception of state and culture is that of an organic whole. As an organic whole, the völkisch state is more than the sum of its parts, and indeed because these parts, called individuals, are fitted together to make a higher unity, within which they in turn become capable of a higher level of life achievement, while also enjoying an enhanced sense

> of security. The individual is bound to this sort of freedom through the fulfillment of his duty in the service of the whole. (Harrington 1996, 176).

Revealingly, Jews were always being compared to rats or other vermin or parasites living off the living globe.

So much for holism being "caring, all-embracing, life sustaining." Note, we are not now swinging entirely the other way, arguing that holism is always wrong and tainted by its associations. That is as extreme as those whom we criticize. Throughout this book, our stance is that metaphors in themselves are neither good nor bad. It is the use that you make of them that counts, and the conclusions. We much doubt that realizing this is going to end debate. Nor should it. Our hope is that we can move the debate into more constructive fields, where there is understanding of ideas outside our paradigm but from which we all can nevertheless learn and move forward.

Chapter Eight

GOD AND THE NEW BIOLOGY

> Science tries to document the factual character of the natural world, and to develop theories that coordinate and explain these facts. Religion, on the other hand, operates in the equally important, but utterly different, realm of human purposes, meanings, and values—subjects that the factual domain of science might illuminate, but can never resolve.
>
> —GOULD 1999, 4

FOUR APPROACHES TO THE SCIENCE-RELIGION RELATIONSHIP

So, what's God got to do with it? Why bring God into a discussion of the New Biology? Even if you are not interested in God, you should at least recognize that others in our society are interested in Him and that it behooves us all to find out what they think about Him, especially if they are having an influence on policy. It is no secret that many do think much about God, and they think "Him" highly relevant to our discussion (Dennett and Winston 2008). Putting things in context, we shall argue that our two metaphors—machine and organism—are absolutely key to the whole God question. There is much to be done before we take the words of Stephen Jay Gould as a given.

As Anglican priest and physical biochemist Arthur Peacocke noted presciently, those who are against God tend to be in the mechanist and reductionist camp. This includes men such as Francis Crick, co-discoverer of the double helix: "the ultimate aim of the modern movement in biology is in fact to explain *all* biology in terms of physics and chemistry" (Peacocke 1986, 1), and ardent Darwinian, Chicago biologist Jerry Coyne: "Religion is but a single brand of superstition (others include beliefs in astrology, paranormal phenomena, homeopathy, and spiritual healing), but it is the most widespread and harmful form of superstition" (Coyne 2015, xii). Then, there are those who are for God. They are more likely to be in the organicist, holist camp. It is little wonder that Arthur Peacocke, he who was the author of the book to which we show our debt by using its title as the title of this chapter, writes: "It is the interconnectedness of the whole biosphere with itself and of the biosphere with all the physical cycles and organisation of planet Earth that have been so urgently manifest to our generation" (Peacocke 1986, 56). Continuing: "We have found that the processes of the world are open-ended and that there are emergent in space-time new organisations of matter-energy which often require epistemologically non-reducible language to expound their distinctiveness."

This is our guiding question: Is the New Biology truly more friendly to religion than the old biology? Since we are now talking specifically of religion and its relationship to science, to aid us on our way, let us introduce a fourfold typology, formulated by the late physicist-theologian Ian Barbour (1997). It is of possible science-religion relationships. It's not the only typology of these relationships, but it's still the most widely used:

1. *Warfare*: Science and religion are competitors and only one can be right.
2. *Independence*: Science and religion talk of different things.
3. *Dialogue*: Science and religion are different, but they speak constructively to each other.
4. *Integration*: Science and religion are or will be one.

There are things to be said for all. Warfare: Clearly, if modern geology is right, then a literal reading of the Bible is wrong about the universality of Noah's Flood. Independence: If God is preparing an eternal home for us, science doesn't really have much to say on this subject, even though some atheists mistakenly hold that science can prove there isn't life after death. Dialogue: If Plato is right, the findings of science show enough design-like phenomena to prove that God exists. Integration: If Teilhard de Chardin is right in what (we shall learn) is his basic claim that evolution leads up to Jesus, this can be combined with information from the fossil record. We do not take these as exclusive divisions. Certainly, on a particular issue—let us say, design and the existence of God—one person might opt for independence and another for dialogue. Perhaps one might even say "independence for me but I don't think you are being unreasonable if you go for dialogue." We take the divisions more as heuristic—helping someone work out what they think or are easy with—than as definitive, where disagreement signals that someone is right and someone is wrong. Indeed, we would say that sometimes people do, and can, combine two approaches. You might think that science conclusively refutes some religious claims—warfare—but that others are simply beyond the scope of science—independence.

We shall take the four possibilities in turn, seeing how they fare under the machine and organism metaphors. Our primary focus will be on the Christian religion—God as loving Creator of heaven and earth, humans as a special creation "made in His image," God coming to earth as Jesus Christ for our well-being, the possibility for humans of eternal salvation with our Creator. As appropriate, we will comment on other religions.

WARFARE

Let us agree that if not open warfare, science and religion can and do clash. Take Noah's Flood. In Chapter 7, we detailed how those working under the machine metaphor devised explanations of the geology of the planet on which we live—Vulcanism, Uniformitarianism, plate tectonics. There is absolutely no way that these could accept or accommodate a deluge that was worldwide and killed off the dinosaurs and whatever. You might think Catastrophism more friendly, and it is true that in the late 1820s, the professor of geology at Oxford University, William Buckland, claimed that the fossil remains of drowned inhabitants of a cave in Yorkshire were evidence of the Noachian deluge (Figure 8.1). He was laughed

FIGURE 8.1 Buckland claimed that the existence of whole skeletons in this cave was evidence that they were washed down by a deluge. Proof of Noah's Flood! (From the revealingly titled *Reliquiae Diluvianae,* 1823.) *Credit:* Reproduced from Willian Buckland, *Reliquiae Diluvianae* (London: J. Murray, 1823).

out of court by all, especially by his fellow Catastrophists. The rocks cannot lie, and they tell not of forty days of worldwide rain and the complete drowning of all animals, save only those lucky enough to sail off in a rather large houseboat.

Not just geology. Biological science including evolutionary science can and certainly does clash with widespread understandings of the Christian religion. You cannot accept modern paleoanthropology—the science of human origins—and accept a literal Adam and Eve story (Ruse 2012). There could not be one single pair, created out of matter, as the origin for the parentage of all humankind. There were bottlenecks in human evolution, but our numbers probably never dropped below about 10,000. Certainly, there was no apple eating to turn us all into flawed sinners. We may be flawed sinners, but not for that reason. Our ancestors' parents were both nice and nasty, and in turn, their parents were both nice and nasty—although if you go back far enough, eventually you come to creatures that were neither nice nor nasty, as they simply didn't have the brains to be either. Nothing very threatening here, you might think. But do not underestimate what taking the Adam and Eve story literally has meant for Christian thinking. St. Augustine asked why it is that God in the form of Jesus had to come to earth and to suffer so dreadfully on the Cross? His answer was that Jesus is a blood sacrifice to atone for our sins—his death is a substitute for our rightful punishment for our transgressions. But why do we merit such punishment if being sinners is simply part of our natural heritage? St. Augustine's answer is that our heritage is not natural. A perfect God made perfect humans. Our sinful nature stems from Adam's sin, abusing the gift of freewill to go against God's explicit command, the effect of which is passed on through his seed so that we all are born with a tendency to sin. Not any old kind of sin—original sin!

There is no getting away from it. Machine-metaphor biology and religion clash. Both cannot be right, hence warfare. In his appropriately named *Faith Versus Fact: Why Science and Religion Are Incompatible,* Jerry Coyne writes:

> Science and religion then are competitors in the business of finding out what is true about our universe. In this goal religion has failed miserably, for its tools for discerning "truth" are useless. These areas are incompatible in precisely the same way, and in the same sense, that rationality is incompatible with irrationality. (Coyne 2015, xvi)

Prima facie, you might think that the organic metaphor is in better shape here if by "better shape" you mean getting along together. Christianity—the other Abrahamic religions also—stresses that humans are unique and more important than other animals:

> 4. what is mankind that you are mindful of them,
> human beings that you care for them?
> 5. You have made them a little lower than the angels
> and crowned them with glory and honor.

6. You made them rulers over the works of your hands;
   you put everything under their feet:
7. all flocks and herds,
   and the animals of the wild,
8. the birds in the sky,
   and the fish in the sea,
   all that swim the paths of the seas. (Ps. 8)

As we have seen, a central thesis of the organic metaphor is that we humans are different from and above other organisms. Whatever the Alpha, we are the Omega. Listen to philosopher John Dupré:

> Though I certainly don't accept that only humans are capable of thought, our forms of consciousness of which we are capable, are very different from those of other terrestrial animals. And human culture, though not unprecedented, involves the articulation and synchronization of a variety of roles and functions that is different in kind from anything else in our experience. (Dupré 2003, 75)

Fair enough. The nasty fly in the ointment for some understandings of religion is that, if possible, organicists are more enthusiastic evolutionists, including humans—especially including humans—than mechanists. Development, change, progress—these are the essence of organicism. The organicist can no more accept Adam and Eve as our literal first parents than the mechanist. And we suspect that the average organicist, at least since the beginning of the nineteenth century, is about as enthusiastic about Noah's Flood as is Richard Dawkins. So much for science and some sorts of religion getting along.

However, interestingly, if you turn to an Eastern religion such as Buddhism, a case might be made for more harmony. Perhaps this is because Western science has always had this parent-child relationship with Christianity—think of how prominently the Genesis notion of the "tree of life" figures in evolutionary reconstructions—whereas Buddhism and Western science have totally different cultural origins. A case can be made for saying that Buddhism is totally indifferent to science as science. It doesn't reject it; it just doesn't think about it. Buddhism, dating from the life and teaching of Gautama Buddha, born and living in Nepal around and after 550 BCE, is an atheistic religion inasmuch as it has no place for a creator god (Harvey 1990). It differs from Christianity also in being committed to the idea of reincarnation—that we have multiple lives in succession (samsara)—and actions and thoughts in this life can have implications for the life that we will live next. The overall aim of life is to break out of this ongoing cycle of existences, achieving "nibbana" (also called "nirvana")—a release from suffering—"dukkha." Nibbana is a kind of nonbeing, although not necessarily nonexistence—it is endless and wholly radiant, the "further shore," the "island amidst the flood," the "highest bliss" (Harvey 1990, 63). All of this takes place against the background of a rather complex ontology. In some versions of Buddhism, there are an infinite number of universes, with galaxies, themselves clustered into 1,000-fold groups. There are innumerable planets, and on them we find inhabitants much like our planet and its denizens. Everything is subject to change, decay, and rebirth—often taking vast quantities of time (eons). Unlike Christianity, in which this world has a beginning, a middle, and an end (the Second Coming), time in Buddhism seems like an endless string, going infinitely back and infinitely forward, but with loops (as time is cyclical—and timeless!) and us somewhere hanging on in the middle.

It seems to us that you could readily be a scientific mechanist or scientific organicist against this background. If you want to believe in evolution, nothing is stopping you. You may fear that mechanism gives you no guarantee that humans will appear on the scene, but given an infinite number of universes with endless time, presumably something such as humans will turn up every now and again, and those worlds without human-like beings are not going to waste. Buddhism posits a kind of hierarchy—somewhat akin to the Christian Great Chain of Being—with lower organisms below us and higher organisms, God-like beings, above us. By being reincarnated as a lower being in other worlds, lots of places for a reincarnated codfish, formerly known as Adolf Hitler, to work out his sins in this world. You might think that because we have this hierarchy, organicism points away from Buddhism. It does not make humans top dogs. But mention of the Great Chain of Being draws attention to the fact that, in this respect, Christianity does not make humans top dogs. God, of course, is at the top, and then there is an entire hierarchy of angels—from seraphim and cherubim at the top, down to archangels and "ordinary" angels at the bottom—above us.

Ending this section, let us acknowledge that we have been playing a bit of a trick on the reader. We are really not suggesting that if you want to go on doing science, doing biology, and if you still want to embrace religion, you need to put on a robe, grab a begging bowl, and forever wander the world chanting Buddhist mantras. Let us not minimize the fact that science and religion can be at loggerheads—at "war," if you like. In the sixteenth century, many Christians were worried that the Copernican system demoted the Earth from its central position, making it but one of a number of planets circling the Sun. Well known is the fact that a century later, Galileo fell afoul of the Church authorities in part because of his heliocentrism. In the nineteenth century, there were certainly clashes on and around Darwin's *Origin*. Listen to his old friend and mentor, Anglican priest and Professor of Geology at Cambridge, Adam Sedgwick:

> If I did not think you a good tempered & truth loving man I should not tell you that, (spite of the great knowledge; store of facts; capital views of the correlations of the various parts of organic nature; admirable hints about the diffusions, thro' wide regions, of nearly related organic beings; &c &c) I have read your book with more pain than pleasure. Parts of it I admired greatly; parts I laughed at till my sides were almost sore; other parts I read with absolute sorrow; because I think them utterly false & grievously mischievous. (Darwin 1985, 7:396)

There was—inevitably with Sedgwick—a great deal more in the same vein. And we have the same kind of thing today. Each of us has spent much of our careers defending Darwin's ideas against the attacks of so-called scientific creationists, better known as biblical literalists or fundamentalists. Ruse was an expert witness in a trial in Arkansas in 1981, testifying (successfully) against a law insisting that if evolution be taught in biology classes in state-funded schools, then so also creationism must be taught (Ruse 1988). Reiss lost his job as Director of Education at the Royal Society because of a misunderstanding over creationism.

But the clashes are far from traditional Christian practice. St. Augustine, around 400 CE, always insisted that although the Bible is true, it is not necessarily literally true. The ancient Jews, he argued, were not educated people like fifth-century Romans, and so God had to speak to them in metaphorical language. As a young man, St. Augustine had been drawn to Manicheanism, which had little time for the Old Testament. He knew the counterarguments and was not about to get trapped. We are made in "God's image," but don't think that that means that God has genitalia and that sort of thing: "the spiritual believers in the Catholic teaching do not believe that God is limited by a bodily shape. When man is said to have been made to the image of God, these words refer to the interior man, where reason and intellect reside" (St. Augustine 1982, 1.5.9).

St. Augustine set the pattern. The question always was not whether science and religion could be harmonized, but whether—as Galileo's critics charged—the cost of harmonization would be too high. It was not until the nineteenth century, in the United States particularly, that the science-religion relationship flared up into outright hostility (Numbers 2006; Hardin et al. 2018).

On the one hand, particularly in the South and among the pioneers moving West, lonely and in need of moral and spiritual support, there was a growth of indigenous Evangelical Protestantism, based on literalistic readings of the Bible—something made possible, thanks to modern mechanized methods of printing, by the increasingly readily available copies of the Good Book (Noll 2002). On the other hand, particularly in the North after the Civil War, there was a counter growth of secular thinking, supposedly more in tune with the growing industrial success of that part of the country. Such highly tendentious books as *History of the Conflict Between Religion and Science* (1875) by J. W. Draper and *History of the Warfare of Science with Theology in Christendom* (1896) by A. D. White appeared on the scene. One senses that these and like writers, every evening, lit a candle to the deity, thanking him for imposing such troubles on Galileo. St. Augustine makes his appearance but perhaps not entirely as the saint might have expected or appreciated. Of scientific thought, we read of "the efforts of Augustine to combat it" (White 1896, 1, 109). Tidbits of Augustinian thinking are scattered liberally. Storms, we learn, are caused by devils. "St Augustine held the same view as beyond controversy." He was in good company. St. Thomas Aquinas (1952), in his "all-authoritative Summa," tells us that "it is a dogma of faith that the demons can produce wind, storms, and rain of fire from heaven" (1, 337). And so it continues down to the present day. The creationists on one side, and the scientists—particularly the New Atheists—on the other. Remember Jerry Coyne. Science and religion are "competitors"; religion has "failed miserably."

Against this, never forget many are not at the extremes but toward the middle. The late-nineteenth-century Anglican theologian Aubrey Moore wrote:

> Darwinism appeared, and, under the guise of a foe, did the work of a friend. It has conferred upon philosophy and religion an inestimable benefit, by showing us that we must choose between two alternatives. Either God is everywhere present in nature, or He is nowhere. He cannot be here, and not there. He cannot delegate his power to demigods called "second causes." In nature everything must be His work or nothing. We must frankly return to the Christian view of direct Divine agency, the immanence of Divine power from end to end, the belief in a God in Whom not only we, but all things have their being, or we must banish him altogether. (A. Moore 1890, 99–100)

Unsurprisingly, Arthur Peacocke—ever eager to see harmony between science and religion—was much given to quoting this very passage.

INDEPENDENCE

The eminent Protestant theologian, the late Langdon Gilkey (1985), stood alongside Ruse in the witness box in Arkansas, speaking up for evolution. Giving a vivid story to support his position, he talked of a trip to the park. One version tells literally about what you did. You made sandwiches, you piled the kids into the car, you played and swam, you got caught in a traffic jam on the way home. One version tells about why you did it. It was mum's birthday, so you and the kids fixed a surprise treat to show your love and appreciation. Two stories—different but no contradictions. Could science and religion be like that? Stephen Jay Gould, quoted at the beginning of this chapter—and fellow witness in Arkansas—thought that they could.

Two stories—different but no contradictions. He argued that there are two "Magisteria"—worldviews—science and religion. Science tells us about, or tries to answer questions about, matters of fact. "Do humans look so much like apes because we share a common ancestor or because creation followed a linear order, with apes representing the step just below us?" "Why does so much of our genetic material (so-called 'junk DNA') serve no apparent function?" (Gould 1999, 55). Religion tells us about, or tries to answer questions about, matters of morality. "Under what conditions (if ever) do we have a right to drive other species to extinction by elimination of their habitats?" Nonoverlapping Magisteria, or NOMA (McGrath 2021).

Most, believers and nonbelievers alike, would say that this is a little too quick and easy. Religion wants to tell us about matters of fact! God exists and created this world. We are made in His image. There is a heaven or a hell awaiting us. Gould recognized this objection and brushed it aside. He was rather sneering about Arthur Peacocke for suggesting that Darwin showed that God's creation is continuous rather than doing all in one moment. Gould asked contemptuously "Is Mr Peacocke's God just retooling himself in the spiffy language of modern science?" (Gould 1999, 217). Without necessarily accepting "Mr Peacocke's God," we do feel that Arthur Peacocke is fully within his rights as a Christian to ask such a question—whatever the answer! So, whether we want to use the language of "Magisteria"—Gould could have given Whitehead a run for his money in making up new terms—let us go at the question a little more carefully, asking first just how one makes the case for the very possibility of independence. Most discussions seem inclined to take the possibility as a given. Is this the best we can do?

We think not. A case can be made—not at once for independence but for its possibility. The secret lies in the nature and role of metaphor. One of Thomas Kuhn's (1962) deepest insights was that paradigms—that, remember, he later identified with metaphors (Kuhn 1993)—work most successfully by putting blinkers on scientists. As in horse races, they work most efficiently if they are not distracted by questions beyond their grasp, or at least are not pertinent to the job in hand.

What relevance does this have to our concerns? Pick up on our metaphors: mechanism, the metaphor of world as a machine; organicism, the metaphor of the world as an organism. However you assign credit, these metaphors have proven hugely successful, and promise to continue to be so. Ask now, then, where is the focus? What is being left out or ignored or suppressed? Let us list four possible candidates (Ruse 2010). We do not claim these are the only ones, and we are empiricists, so we don't want to say absolutely that nothing could change. We ask, where are we at this moment?

Existence

The first issue is that of existence. You might say that surely today's science deals with that and provides the answer, namely the Big Bang. The question we are asking, however, is not so much a straightforward, efficient-cause question, but more one at a kind of metaphysical level. Why is there anything at all? Why not nothing—not empty space but nothing existing? It is what the German philosopher Martin Heidegger (1959) called the "fundamental question of metaphysics." Some philosophers, notably the famous Austro-British thinker in the first half of the last century, Ludwig Wittgenstein (1965), thought this was a bogus question. You can never answer it. We, to the contrary, don't think it is silly at all. And "never" is a long time. The point we want to make is that because of the machine metaphor, science doesn't even set out to answer it because it doesn't—it cannot—ask it. Machines are like the cookbook—first, take your hare. Of course, we can ask where the steel from the automobile came from, but you just keep going back, and it becomes irrelevant to the running of the car—or the world. It just is.

What does science under the organic metaphor have to say about all of this? We don't see that it would say much different. Organisms have to come from somewhere, and the answer obviously is from the seeds of earlier organisms. The oak tree grows from the acorn, and the acorn was produced by an earlier existing oak tree. You can keep going back, but like the woman said, "It's turtles all the way down." You cannot, within the bounds of organismic science, say why there are seeds in the first place. Or rather, if you are an evolutionist, presumably you can say why the oak tree went in for acorns rather than, say, insisting on grafts or some such thing to propagate. But if you keep going back, you get to those proto-organisms—monads—at the beginning of life, and then you just keep going back with the preorganic. And it's no good appealing to the Big Bang or anything like that because then the question becomes "Why the Big Bang?" and then you say either that there was something before the Big Bang—a Big Contraction perhaps—or there was nothing. Precisely! Why is there something rather than nothing?

Here, of course, is the point at which religion steps forward and says, "Let me help." There is something rather than nothing because it is the creation of a good God, which obviously then brings up the question posed by first-year undergraduates—and Richard Dawkins. What caused God? To the surprise of undergraduates, philosophers and theologians and Christians generally have thought about this one. The answer is that God is a necessary being and thus in some sense cause of Himself. Just as it is meaningless to ask when did two plus two first equal four, so it is meaningless to ask why God exists—or rather that it is not so much meaningless as not needed. God just is as two plus two equals four just is. Now, as you might well imagine, this has led to a huge amount of discussion. Is it possible to have a logically necessary being? Anselm's ontological argument purports to show the necessary existence of God on the grounds that since God is by definition that which no greater can be conceived, if God does not exist, then something greater can be conceived, namely an existent God. A contradiction showing that God must exist, logically necessarily.

Many, including David Hume, maintain that this is not a convincing argument:

> Nothing is demonstrable, unless the contrary is a contradiction. Nothing, that is directly conceivable, implies a contradiction. Whatever we conceive as existent, we can also conceive as non-existent. There is no being, therefore, whose non-existence implies a contradiction. Consequently there is no Being whose contradiction is demonstrable. (Hume [1779] 1990, 189)

Countering this, many philosophers, notably Aquinas, argue that God's necessary existence is not logically necessary existence but rather some kind of existence where God is cause of Himself—"aseity." Fortunately, we need not pursue this question further. Our point is made. Science cannot speak to the question. Religion can and does. Whether it does so adequately is another matter. Expectedly, Richard Dawkins is unimpressed. In one pithy sentence, he manages to put the boot into both the ontological argument and philosophers. "I mean it as a compliment when I say that you could almost define a philosopher as someone who won't take common sense as an answer" (Dawkins 2006, 83). As they say, with friends like this . . . Leave it at that.

However, before moving on, it is worth noting that Buddhism would leave the fundamental question unasked. There is no creator god. So, for the Buddhist, the fact of existence is simply an unexplained given, as it is for the scientist. All existence is dependent on the existence of others. "All beings and phenomena are caused to exist by other beings and phenomena, and are dependent on them. Further, the beings and phenomena thus caused to exist also cause other beings and phenomena to exist. Things and beings perpetually arise and perpetually cease because other things and beings perpetually arise and perpetually cease" (O'Brien 2019). But that is as far as you can go:

> In Buddhism, unlike other religious philosophies, there is no teaching of a First Cause. How all this arising and ceasing began—or even if it had a beginning—is not discussed, contemplated or explained. The Buddha emphasized understanding the nature of things as-they-are rather than speculating on what might have happened in the past or what might happen in the future. (O'Brien 2019)

Science and religion are independent because there is nothing to quarrel about.

Values

Stephen Jay Gould is right that religion talks of values—talking now of moral values and not of aesthetic and like values. Where Gould was wrong is to think that religion talks only of values. What about the other side of NOMA? Does science eschew value talk? We have talked about this already and how the machine metaphor welcomes—demands—a David Hume–type answer. Machines in themselves just are; they are neither good nor bad. Of course, some machines are designed in such a way that they really have only one use, and we can judge that good or bad. But even here, it is complex. A guillotine is really only for chopping off heads. We suppose one could use it to top and tail large carrots, but it is not truly built for that. We would say, emphatically, that the machine is not a good thing. When Sophie Scholl of the White Rose group was guillotined for passing out anti-Nazi literature, it was a very bad thing indeed. But what about Hitler? A lot of people would say that it would have been a very good thing had he been guillotined long before he committed suicide. It does seem that we ascribe goodness and badness to machines rather than find it there. In other words, science doesn't justify morality because—thanks to its root metaphor—it is not really about morality at all.

At least, it is not about morality in the sense of justifying morality. Telling you why it is wrong to mark up library books. Darwinians would argue strongly that their theory tells you a lot about why people think and act morally, or not. Most obvious, to recap from Chapter 3, is what is known as "reciprocal altruism" (Trivers 1971). You scratch my back, and I will scratch yours. Remember, Darwin wrote: "as the reasoning powers and foresight of the members [of a tribe] became improved, each man would soon learn from experience that if he aided his fellow-men, he would commonly receive aid in return" (Darwin 1871, 1, 163). Natural selection promotes such thinking and behavior for its adaptive value. No justification involved. If for some reason humans couldn't interact with one another, reciprocal altruism is a bit of a waste of time. As has been put provocatively, "morality is an illusion put in place by our genes to make us good cooperators" (Ruse & Wilson 1985, 108; see also Ruse & Wilson 1986). Morality is real. It is the belief that it is justified—existing irrelevant of us, like two plus two equals four—that is illusory.

What about moral values under the organism metaphor? Here, things become more complex. The very essence of organicism is that things—organisms particularly—do have value and that value increases as we approach humankind. Remember Brian Goodwin (King 1996). We have got to escape the Darwinian metaphors of "competition and conflict and survival," replacing them with metaphors stressing organisms as "co-operative as they are competitive." We must turn from "nature red in tooth and claw, with fierce competition and the survivors coming away with the spoils." We need a new perspective where the "whole metaphor of evolution, instead of being one of competition, conflict and survival, becomes one of creativity and transformation." Other organicists think along similar lines. To think that nature is without value is simply wrong. But what, then, does this imply for our discussion? Not that the organicists are wrong and the mechanists right, or conversely. Rather, that whereas independence from religion is an open possibility for the mechanist, perhaps indeed is the preferred option for the mechanist, it is not really a possibility for the organicist. They are directed to dialogue or integration. Note that we are not saying that the mechanist denies that religion can speak to moral issues. The whole point is that the mechanist agrees that religion can speak to moral issues, although the mechanist might also argue that neither science nor religion speaks fully to morality, and we need some kind of secular approach—G. E. Moore's (1903) nonnatural properties or some such thing. What we are saying is that whereas the organicist's science can speak to moral issues, the mechanist's science cannot speak to such issues. Independence.

Consciousness

Third, the consciousness question. At one level, the message from the mechanist is blunt. We are not in the consciousness-explaining business. Leibniz, in his *Monadology,* put his finger right on it. Machines don't think. At least, machines are not conscious. And, you won't be surprised at this, science cannot answer the consciousness question. Leibniz was right. You describe a machine, and when you have finished, there is no hint of consciousness. You are working in the Cartesian world of material substance, *res extensa*. No consciousness in; no consciousness out.

> One is obliged to admit that perception and what depends upon it is inexplicable on mechanical principles, that is, by figures and motions. In imagining that there is a machine whose construction would enable it to think, to sense, and to have perception, one could conceive it enlarged while retaining the same proportions, so that one could enter into it, just like into a windmill. Supposing this, one should, when visiting within it, find only parts pushing one another, and never anything by which to explain a perception. Thus it is in the simple substance, and not in the composite or in the machine, that one must look for perception. (Leibniz 1714, 215)

Now, notice carefully what is being said here and what is not being said. Most importantly, no one is saying that the mechanist can say nothing interesting about the mind and its relationship to body. Most obviously, one can point to physical features or events that provoke reactions in the mind and conversely. You can, for instance, locate the part of the brain that controls language ability, and you can show that thinking certain thoughts provokes bodily responses. This is not explaining what consciousness is, but it is throwing light on how it operates and what is its value. Second, one can as a mechanist make some pertinent speculations about the nature of consciousness—not explaining it but showing the domains or possibilities open. The mathematician-philosopher William Kingdom Clifford, writing in the time of Darwin, thought the theory of evolution through natural selection very suggestive:

> We cannot suppose that so enormous a jump from one creature to another should have occurred at any point in the process of evolution as the introduction of a fact entirely different and absolutely separate from the physical fact. It is impossible for anybody to point out the particular place in the line of descent where that event can be supposed to have taken place. The only thing that we can come to, if we accept the doctrine of evolution at all, is that even in the very lowest organism, even in the Amoeba which swims about in our own blood, there is something or other, inconceivably simple to us, which is of the same nature with our own consciousness, although not of the same complexity. (Clifford [1874] 1901, 38–39)

What Clifford is suggesting is that Darwinism—machine-metaphor Darwinism—points us to what is known as "panpsychic monism," a thesis associated with Spinoza that mind and matter are in some sense one. There is no separate *res extensa* and *res cogitans*. The stuff of the universe is both mind and matter. As critics point out, in some sense, you are suggesting that molecules think. And if that stretches the imagination, what of the combination problem? If you put a bunch of molecules together, why don't they all think separately (Leidenhag 2019)? Why do they come together as one thinking being or another—to name two men born on the same day, Charles Darwin or Abraham Lincoln? Against this, thanks to physics, no one today believes that matter is the kind of stuff talked of by the atomists, or Descartes for that matter. Complementarity—wave or particle—put an end to that. As do proven phenomena such as quantum entanglement—information is passed at once across vast differences, as something happening to a particle in one part of the universe is reflected in something happening to another particle in another part of the universe. These are not isolated Democritean atoms at work and play.

Our job here is to show how the mechanist might deal with consciousness. To be honest, we are not sure how the organicist is going to say anything that is fundamentally different. If you think that the world is organic, this does not at once tell you what consciousness is or its relationship to body. However, as has been pointed out by many prominent organicists—Whitehead most notably—panpsychic monism seems almost expressly designed for organicism. The world has value, in some sense it is living, and so naturally one thinks of mind as being all-pervasive. "The doctrine that I am maintaining is that neither physical nature nor life can be understood unless we fuse them together as essential factors in the composition of 'really real' things whose interconnections and individual characters constitute the universe" (Whitehead 1938, 205). Continuing: "this sharp division between mentality and nature has no ground in our fundamental observation. We find ourselves living within nature." Hence: "I conclude that we should conceive mental operations as among the factors which make up the constitution of nature" (214).

What of religion on this topic? Judeo-Christianity tells us:

> 26 Then God said, "Let us make mankind in our image, in our likeness, so that they may rule over the fish in the sea and the birds in the sky, over the livestock and all the wild animals, and over all the creatures that move along the ground."
>
> 27 So God created mankind in his own image, in the image of God he created them; male and female he created them. (Gen. 1)

This is not taken in a physical sense—remember, St. Augustine stressed that God does not have both male and female physical features—rather, in some kind of mental sense. We have intelligence—we can explore the world God created—and we have a sense of morality, of what is right and what is wrong. Now, whatever you might say about science saying why we have intelligence or of the kind we have, same with morality—we shall speak more of this in a moment—really the claim that God did it all is neither here nor there for either the mechanist or the organicist. We can surely grant independence here. (Note we are saying we can surely grant the option of independence. This is not to stop someone saying they want a stronger link like dialogue.)

Purpose

The most obvious thing you ask about a machine is "what's it for?" As we have seen, however, the most important thing about the world-as-a-machine metaphor is that those teleological questions are dropped. Why the moon exists is not a scientific question, even though as a Christian you might think that God created the moon in order to light the way home for drunken philosophers. The early mechanists found that teleological questions didn't help at all in inquiry, and so they were cast aside. God became a retired engineer. The focus was on, and entirely on, efficient causes and working according to natural law. On and on, ever the same, indefinitely. A clock that never stops. Equally, one where no one ever tells time, just looks at the workings—ticktock, ticktock. Organisms were a problem, but Darwin dealt with that one. Organisms, including humans, are machines. In the world of science, without meaning or purpose. True, Darwin happily used final-cause thinking, but the very point of his theory was that he showed final-cause thinking could be explained by efficient-cause thinking.

Steven Weinberg (1977) said, "The more the universe seems comprehensible, the more it also seems pointless" (154). Why are we not surprised here? As we have just seen, the machine metaphor of science doesn't allow those sorts of questions. That is not how the metaphor functions. There is no point to the universe of science because—qua science—you are simply not asking or allowed to ask questions about purposes and ends. What of the world-as-an-organism metaphor? In the case of mechanism, the answer is easy to give. Mechanism has no talk of ends—no talk of ultimate ends, that is—so it says nothing about religious claims about ends. Organicism, however, does talk of ends—humans! So, to answer the question here, ask the supplementary question about what the religious claims are.

In the case of Christianity, the ultimate intended end is an eternity of bliss with the Creator. Organicism being science is certainly not going to tell you about eternities, with or without bliss with the Creator. It is not going to deny them either, so in this respect you can claim independence. But as with the earlier areas—value and consciousness—the answer requires a bit more. The Christian invites you to think teleologically—the Buddhist too if you are thinking about nibbana—so a case could be made that organicism is, as one might say, more user friendly toward the Christian than the mechanist. Perhaps the relationship between science here in this sense and religion in this sense is a little stronger. We shall see about this in a moment. Parenthetically, traditional Judaism didn't have much time for the afterlife. So, you might say the organicist approach is no more relevant than the mechanist approach. This, however, is not quite true, since traditional Judaism is very focused on the successful continuation of the Jewish people, and that surely is teleological.

DIALOGUE

Move on now to the third of Barbour's options, dialogue. Here, science and religion are separate but in respects can speak constructively to one another. As with independence, we take up the four topics or areas.

Existence

Why is there something rather than nothing? As we have seen, in an important sense neither the machine-metaphor approach nor the organism-metaphor approach has anything to say on this score. So, independence certainly seems like the obvious option. But could there be another option? Could the person of religion turn to science to support the claim that there is something rather than nothing because of the creative act of a loving God? They certainly could and do. So, in this sense, for those so inclined, dialogue seems an option. Note, not the mandated Christian option. There are those in the Christian tradition, such as Søren Kierkegaard in the nineteenth century and Karl Barth in the twentieth century, and those in other faiths, who want nothing to do with using science to support or to flesh out religious claims—nothing to do with what is usually known as "natural theology"—and want to put all religious claims down to faith—"revealed theology." But what of the many Christians such as most Catholics and Anglicans—and those of other faiths—who do want to appeal to science for support? Or critics who want to appeal to science for refutation?

Start with the positive case. Most obviously, the proofs of God's existence and, since we are dealing with science, the empirically based proofs of God's existence. In the case of some, it is really a bit of a stalemate. With respect to miracles, for example, scientists—mechanists and organicists—have committed to what is known as "methodological naturalism"—nature can be explained by unbroken law—so they are simply not going to allow miracles in their day-to-day work. Of course, they argue that their stance shows that appealing to miracles is truly unnecessary. Take the miracle supporting the sainthood of John Paul II:

> The late pontiff was credited with curing Floribeth Mora, a woman from the town who had a severe brain injury. Her family prayed to the pope's memory and says she was cured on May 1, 2011.
>
> Mora's neighbor Cecilia Chavez voiced the community's feelings.
>
> "How can it be that in a small country such as Costa Rica, in this poor small neighborhood, this miracle took place? It's amazing! There are no words to describe it," she said.
>
> Floribeth Mora had walked into a hospital in Costa Rica's capital San Jose complaining of a headache.
>
> Neurosurgeon Alejandro Vargas Roman, who diagnosed her with a brain aneurysm, says the question of why it disappeared without surgical intervention is without explanation. "I have never read about this anywhere around the world," he said. (Ridgwell 2013)

Well, yes. In the light of something like this, the kindest thing to say is that if you want to believe in miracles, go ahead and do so, but don't pretend that science is behind you. Even if we accept that science does not now give an answer (as to why this event was natural), this is no proof that science can never give an answer. And past experience very much suggests that in the future, science will be able to give an answer.

The most pertinent of the empirical arguments is obviously the argument from design, the teleological argument. For the Darwinian mechanist, that no longer works. You do not need to invoke God as an explanation of design-like adaptations. Natural selection does the job for you. This does not mean that all hopes of a dialogue are now cast aside. If you take the attitude of John Henry Newman (1973)—"I believe in design because I believe in God, not in God because I believe in design" (97)—then Darwinian science can certainly flesh out your faith commitment to God's nature. As an undergraduate, Darwin learnt that from the classic text by Kirby and Spence: *An Introduction to Entomology: or Elements of the Natural History of Insects* (1815–1828). Through nature, one gets to God: "no study affords a fairer opportunity of leading the young mind by a natural and pleasing path to the great truths of Religion, and of impressing it with the most lively ideas of the power, wisdom, and goodness of the Creator" (35). This is pertinent also for the organicist. Because one is an organicist, one is not necessarily a believer. Thomas Nagel is openly atheistic. Almost paradoxically, as nonbelievers in the overpowering efficacy of natural selection, there is a tendency among organicists to downplay adaptation and so, in respects, for such a thinker, nature is that much less able to illustrate the power and goodness of God. Inasmuch as the organicist does see adaptation, whether or not powered by natural selection, the case is the same as that for mechanism. Dialogue but not proof.

The main spoiler to all this happy intercourse between scientists and the religious is the problem of evil. Does one say there is indeed a dialogue between science and religion, and science shows religion false?! To some, it seems that way. In a letter to his friend Asa Gray, Darwin worried that this might be the case. "I cannot persuade myself that a beneficent & omnipotent God would have designedly created the Ichneumonidæ with the express intention of their feeding within the living bodies of caterpillars, or that a cat should play with mice" (Letter to Asa Gray, 22, May 1860, in Darwin 1985–, 8, 224). Interestingly, however, Darwin's theory can speak to some of the worries here. For a start, the caterpillars parasitized by ichneumonid wasps almost certainly don't suffer—lacking the mental capacity to do so. Then, there is the more general problem of natural evil—the Lisbon earthquake and cats playing with mice. As Leibniz pointed out, one is not limiting God's powers by pointing out that He cannot do the impossible. Of all people, Richard Dawkins (1983) himself floats the possibility that natural selection is the only natural way of getting design-like features in the world. In other words, pain and suffering are part of a package deal to get functioning organisms. We are not sure how appreciative Dawkins would be to learn that he did not get to this point first: "lions would not thrive unless asses were killed" (Aquinas 1952, 1a, 25, 6). On the other hand, there is moral evil. Heinrich Himmler. The usual way that this is tackled by the Christian is to argue this is a function of freewill. It was better that God create free creatures—that is, creatures in His image—than make us blind robots or unthinking brutes. "God therefore neither wills evil to be done, nor wills it not to be done, but he will to permit evil to be done, and this is good" (Aquinas 1952, 1a, 82, 1). Here, the Darwinian points out that one of our most important adaptations is the power to make choices. Everything is not predetermined. Think of missiles. There are the simple ones that when shot off cannot change track—let us say the ants—and then there are the complex ones that can respond to changes by the target—humans. We are like one of the Mars Rovers (Dennett 1984). When it meets an obstacle, it does not have to be controlled from base like a marionette. It can negotiate its own way round. Again, it seems that Darwinism comes to the rescue.

Is the organicist going to argue otherwise? Probably not, although our suspicion is that the organicist may feel less troubled by the problem of natural evil than the mechanist. The organicist is not going to deny the troubles caused by the Lisbon earthquake but might be more inclined to downplay the worries that Darwin had about the suffering caused by the evolutionary process itself. The main force for change is internal rather than the external force of natural selection. So, there is simply not going to be the wholesale carnage of Darwinism. Remember John Muir who claimed never to have seen blood and guts out in nature thanks to the struggle for existence. Neither Christianity nor any of the other major religions, including Buddhism, wants to deny the problem of natural evil. Indeed, for the Christian certainly, and as we have seen for Buddhism, this is very much part of life's journey through this world—what the poet

Keats called the "vale of soul making." But inasmuch as the Creator is seen as less harsh than the God of Darwinism, this is surely a note of favor for organicism. This doesn't prove Christianity—or any other religion—true, but it does help. And that surely is what dialogue is all about.

Values

Note first, when we say that the machinemetaphor is drained of values, this means that it cannot justify values. As we know, it can and does say a lot about values at the substantive or normative level—the values we have and how they have been shaped by natural selection as well as culture. It can tell us why we think that values are justified—why we think there is a metaethical foundation. If we didn't think this, more people would cheat, and the normative value system would break down. But it cannot itself offer a real foundation (Ruse 2021b). The organic model is different in this respect. It does seem to have values built right in. It is genuinely teleological through and through. The plates on the back of stegosaurus have value because they led to temperature control in the prehistoric brute. A flower has value because it attracts a pollinating insect. Our brains have value because they let us navigate / manipulate and live in the world. Perhaps there is value in the whole system of growth. A mighty oak has greater value than its acorn. Humans have greater value than warthogs, and warthogs have greater value than pond slime. If you are into the Gaia hypothesis, then the whole system has value. Don't desecrate Mother Earth!

So, if it is values you want, then you seem to be better off with the organic model or metaphor. The mechanist will respond that no one is saying that the machine model has no values. It is all a matter of whether they are justified—whether, ultimately, they are all a question of what humans need rather than what humans ought, in some objective way, to have. In fact, one can think of scenarios where one might morally prefer the mechanist side. Suppose there is some rare plant that has medicinal virtues for an uncommon but very unpleasant and dangerous childhood complaint. By harvesting about half of the plant—which can only grow in the wild—one can provide just enough to treat all the afflicted children.

Enough of the plant remains that it can keep up supplies year after year. But then it turns out that some rare butterfly feeds on and only on the plant. For some reason, to keep up the population, you need most of the plants. Too few and the butterfly no longer has a critical number of individuals and goes extinct. What should you do? We suspect that the mechanist will regret the butterflies but value the children more, whereas some organicists might think the rights of the butterflies to be absolute. "A life-centered system of environmental ethics is opposed to human-centered ones precisely on this point. From the perspective of a life-centered theory, we have prima facie moral obligations that are owed to wild plants and animals themselves as members of the Earth's biotic community" (Taylor 1981, 197). We stress, as always, that we the authors are not saying or arguing for anything definitive. We are saying that, as always, things are a little more complex than first impressions suggest.

Things get even more complex when we turn to religion. You might think that Christianity will at once opt for the warm and friendly organic model over the harsh and forbidding mechanist model. Not necessarily! For a start, positively in favor of the mechanist model, Catholic natural law theory starts with the world as lawlike (Ruse 2015). God's will, His imposition of moral norms, is not arbitrary but natural. He follows the way the world is made. It is natural to love your children and cherish them, and hence morally acceptable, nay obligatory. It is not natural to hate and kill your children, and hence it is not morally acceptable. Darwinism fits happily with this. Loving your kids is a number 1 moral obligation. Killing your kids is a number 1 no-no. We are not saying that everything in Catholic natural law theory fits happily with everything in Darwinian evolutionary biology—one could well imagine differences over sexual behavior—but they are talking in the same ballpark. Darwinians may not feel the need for a moral foundation, but if Catholics want to give one, no one is stopping them. Darwinism and hence the mechanical in this respect gives the Christians a piece of candy.

For a second point, negatively against the organic model, a large number of evangelical Christians believe that the world was created for humans, and we have the right to do what we will with the world. God will always be there to put things right. No nonsense about values out there in the world to be discovered. "And God blessed them, and God said unto them, Be fruitful, and multiply, and replenish the earth, and subdue it: and have dominion over the fish of the sea, and over the fowl of the air, and over every living thing that moveth upon the earth" (Gen. 1:28). The butterflies go extinct? *Tant pis*! In 2005, the conservative evangelical Cornwall Alliance for the Stewardship of Creation stated that "just as good engineers build multiple layers of protection into complex buildings and systems, so also the wise Creator has built multiple self-protecting and self-correcting layers into His world," adding: "Therefore a Biblical theology of Earth stewardship will recognize the superintending hand of God protecting the Earth. Particularly when it is combined with . . . observations about the resiliency of the Earth because of God's wise design, this ought to make Christians inherently skeptical of claims that this or that human action threatens permanent and catastrophic damage to the Earth" (Larson & Ruse 2017, 253–254). Global warming? "What, me worry?"

Not all Christians feel this way. Thomas Berry, a Catholic priest no less, had a theological vision of the world that made the organic thesis central. "The universe is not a vast smudge of matter, some jellylike substance extended indefinitely in space. Nor is the universe a collection of unrelated particles. The universe is, rather, a vast multiplicity of individual realities with both qualitative and quantitative differences, all in the spiritual-physical community with one another" (Berry 2009, 71–72). Integration, and probably dialogue, push us to the Earth-as-an-organism. "This unique mode of Earth-being is expressed primarily in the number and diversity of living forms that exist on Earth, living forms so integral to one another and with the structure and functioning of the planet that we can appropriately speak of Earth as a 'Living Planet'" (110). The best surprise is no surprise. What about a bit of Darwin bashing? "Darwin had only a minimal awareness of the cooperative and mutual dependence of each form of life on the other forms of life. This is remarkable: he himself discovered the great web of life, yet he did not have a full appreciation of the principle of intercommunion" (73).

Just a lowly priest. A buck private in the army of Roman Catholicism. What did the commander in chief have to say? Go back to the 2014 encyclical of Pope Francis (2015). He tells us that we are called upon "to accept the world as a sacrament of communion, as a way of sharing with God and our neighbours on a global scale. It is our humble conviction that the divine and the human meet in the slightest detail in the seamless garment of God's creation, in the last speck of dust of our planet." Continuing: "We are not God. The earth was here before us and it has been given to us. This allows us to respond to the charge that Judaeo-Christian thinking, on the basis of the Genesis account which grants man 'dominion' over the earth (cf. Gen 1:28), has encouraged the unbridled exploitation of nature by painting him as domineering and destructive by nature. This is not a correct interpretation of the Bible as understood by the Church." Note, however, it is built into Francis's thinking that the world and its contents do not have value in their own right. They have it because they were created by God, and He gave them value. This points us to some important issues about consciousness. So, let us pick right up on them.

Consciousness

Already, in the discussion of consciousness in the section on independence above, we have raised issues pertinent to dialogue. And the worries that these issues raise are intensified when we discuss values as we have just above. If you take up the organic model, then you are into values. The burning question is what is the ultimate source of these values? Christians—not just in the extreme evangelical corner of the religion—worry that ideas such as the Gaia hypothesis point to heresy. For many Christians, ultimately, the earth and its contents must in themselves be value neutral. They can and do have value, but it cannot be self-generated value. Thus, the Christian head of biology at Wheaton College, Billy Graham's alma mater: "Scripture provides a logical value system. It establishes that the whole creation in general, and every part of it in particular, has a value given to it by God. This does not mean that the creation is inherently good or that it has the right to exist on its own merits, independent of God. Its goodness is derived from its Creator and so is a kind of "grace" goodness, freely given in love, not grudgingly merited by right" (Van Dyke et al. 1996, 53). Likewise, the Lutherans: "The earth is very good. Neither demonic nor divine, neither meaningless nor sufficient unto itself, it receives its meaning and value from God" (Evangelical Lutheran Church in America 1996, 245). This means that theologically, one must be very wary of ideas such as the Gaia hypothesis. "Though the hypothesis itself can be considered reasonably scientific, it has spawned a host of ideas and philosophies which reach out to deify the earth" (Van Dyke et al. 1996, 139). This is idolatry. It would be unscriptural to worship the earth. "Creation worships the Creator" (ELCA 1996, 244).

The worries about consciousness are now starting to come out. If you get into a world of values in themselves—and panpsychic monism certainly seems to point that way—then you are starting to edge dangerously close to pantheism, God and the world are one. Spinoza again! *Deus sive natura.* Take the Mormons, of all people. You might expect these would be the last people to endorse world pictures and metaphors that, given the value-impregnated nature of the world, virtually beg for environmentalism. Overall, if there is a right-wing position looking for support, go no further than the State of Utah. Surely the world has no standing in its own right and is just there for our use? And yet, that is the trouble with revelations. They link you up with some unexpected traveling companions. *The Book of Moses* (dictated to Joseph Smith in 1830 and 1831 and now incorporated in *The Pearl of Great Price,* one of the four sacred texts of the faith) shows that the world itself is quite able to express fairly strong emotions: "Wo, wo is me, the mother of men; I am brained, I am weary, because of the wickedness of my children. When shall I crest, and be cleansed from the filthiness which is gone forth out of me? When will my Creator sanctify me, that I may rest, and righteousness for a season abide upon my face?" (7:48). The talk is of a creator, but the world is more now than just of value. It is a conscious being of value. We are certainly starting to edge in the direction of pantheism.

King Charles III of the UK is one who goes in this direction of turning the earth into an actual being. He quotes, if somewhat gnomically, the alchemic "Emerald Tablet of Hermes": "And as all things are One, so all things have their birth from this One Thing by adaptation. Its power is integrating if it be turned into Earth" (Prince of Wales et al. 2010, 120). Perhaps enough of the Mormons are Anglophiles; they are happy to be at one with the new King of England. One doubts they would be quite so happy if they went the other way to California, home of many of the pagans. One distinctive aspect of paganism is that they are happy to welcome to their folds people with a wide range of beliefs—beliefs that for Christians would lead at least to another Thirty Years War. One thing does unite them. If you are looking for a group who are 100 percent for the organic model and 100 percent against the mechanical model, your quest is over. "You and I and every other person creature, tree, and flower are cells in the greater living body of Mother Earth—or Gaea, as many call Her. And the living Earth is only one of countless bodies of all sizes—planets, moons, asteroids, comets, meteorites, planetoids, and planetesimals—that make up our solar system" (Zell-Ravenheart 2004, 53). Somewhat humorously, many of today's leading pagans had good American liberal arts colleges' educations. In a previous incarnation, Tim Zell graduated from Westminster College in Fulton, Missouri. They are well grounded in the classics. Back to Plato! "All of it—every atom in your body; you and all your family, friends, and neighbors throughout the world; every living creature and plant upon the face of the earth; every planet, moon, and comet in the solar system; every star in the Milky Way galaxy, every galaxy in the vast and infinite universe—all are connected into the one great Web of Unity, one great Universal 'Internet' of Space and Time, Matter, and Energy" (Zell-Ravenheart 2004, 52).

Leave matters there. If you take dialogue seriously, the respective attitudes toward values of our rival root metaphors do get mixed up in religious responses. However, far from the obvious connections—mechanism religiously bad, organicism religiously good—being well taken, when you start to dig into issues such as consciousness, in respects, things even get reversed. It is all a little bit like Christian on the straight and narrow to the Celestial City. On both sides, there are lots of tempting paths leading to dangerous heretical sloughs. Pantheism is one of the most tempting.

Purpose

We have already taken up, in detail in the independence section, the question of purpose and ends. Next, in the integration section, purpose will play a major role. The conclusion of the independence section was simple and clear. Mechanism—Darwinian mechanism—simply doesn't have a dog in this fight. Purpose, that is, that is beyond mechanistic efficient causation. Darwin's genius was to show that we can explain final cause mechanistically. This is not the ultimate purpose of the Platonist or Aristotelian. Moreover, as we saw earlier in this chapter, while the Christian or anyone else can certainly impose this kind of purpose on the picture, Darwinism is certainly not going to help. The essence of Darwinian evolution is non-directionality. That certainly seems to pose a problem for dialogue, or at least suggest that dialogue is going to end unhappily—for the mechanist, that is. Humans might not be necessary beings, like God, but for the Christian, their existence cannot be just contingent, a matter of luck. Humans had to appear here on earth. Presumably, they might have had bright blue skin and perhaps six fingers. Perhaps they might have had features that many Christians think disgusting and immoral. When you think of the ways some organisms reproduce here on earth—there are species of all-female fish, reproducing asexually, but needing the males of a different species to kick things into action—it is not totally implausible that our possible humans might have needed a romp with bonobos in order for the sex organs to function properly. However, thinking beings with a moral sense are not a matter of choice. Unfortunately, Darwinism simply doesn't guarantee that humans would have evolved. We did, but we didn't have to, nor candidly would we have been expected to. Making reference to the asteroid sixty-five million years ago that wiped out so much life and made possible the rise of the mammals, Stephen Jay Gould joked: "Since dinosaurs were not moving toward markedly larger brains, and since such a prospect may lie outside the capabilities of reptilian design (Jerison 1973; Hopson 1977), we must assume that consciousness would not have evolved on our planet if a cosmic catastrophe had not claimed the dinosaurs as victims. In an entirely literal sense, we owe our existence, as large and reasoning mammals, to our lucky stars" (Gould 1988, 318).

Organicism, with its central belief that humans are special and the whole of evolution leads up to them, seems to be the *saveur du jour*. Any hope for the mechanist? We suppose that if you believe in infinite multiverses, human-like beings were bound to evolve an infinite number of times. But so also would an infinite number of epsilon minuses, who might be good in bed but hardly philosopher kings. We suppose God could consider the options and only actualize that or those that do lead to human-like beings. This might work if you don't really think this is cheating and thus unworthy of God. It assumes of course that God has full knowledge of the future or possible future. In this age of quantum uncertainty, is that going to be credible? Perhaps you might cut the knot by going in a different direction, suggesting that—appearances notwithstanding—Darwinian mechanism can and does have direction. This certainly seems to be the position of that ultra-Darwinian Richard Dawkins. Remember from Chapter 3: "Directionalist common sense surely wins on the very long time scale: once there was only blue-green slime and now there are sharp-eyed metazoa" (Dawkins & Krebs 1979, 508).

How did this all come about? Spell out Dawkins's reasoning in a little more detail. Those sharp-eyed metazoa, aka organisms with the largest on-board computers—human beings—got where they are in the same way that computers were developed, as sophisticated weaponry. Competing nations get into "arms races," and thus features get ever improved. Dawkins refers to a notion known as an animal's EQ, standing for "encephalization quotient" (Jerison 1973)—a kind of cross-species IQ measurement that takes into account the size of the organism. Whales require much bigger brains than shrews simply because they need more computing power to get their bigger bodies to function. With the surplus left over, one can then scale raw intelligence. "The fact that humans have an EQ of 7 and hippos an EQ of 0.3 may not literally mean that humans are 23 times as clever as hippos! But the EQ as measured is probably telling us *something* about how much 'computing power' an animal has in its head, over and above the irreducible amount of computing power needed for the routine running of its large or small body" (Dawkins 1986, 189). We won!

There are other suggestions as to how you might get progress to humans in a Darwinism fashion. British paleontologist Simon Conway Morris (2003) argues that only certain areas of potential morphological space will be able to support functional life. This puts constraints on the direction of evolution. Not all pathways are open. Those that are open tend to be used again and again—"convergence." Conway Morris concludes that that the historical course of nature is not random but strongly selection-constrained along certain pathways and to certain destinations. And our own very existence shows that a kind of cultural adaptive niche exists—a niche that prizes intelligence and social abilities. We grabbed it! Humans came into their own. "We may be unique, but paradoxically those properties that define our uniqueness can still be inherent in the evolutionary process. In other words, if we humans had not evolved then something more-or-less identical would have emerged sooner or later" (Conway Morris 2003, 196).

Expectedly, not everyone—even Darwinians—is overly enthused by these suggestions. Arms races come up against the immortal words of the distinguished paleontologist, the late Jack Sepkoski. "I see intelligence as just one of a variety of adaptations among tetrapods for survival. Running fast in a herd while being as dumb as shit, I think, is a very good adaptation for survival" (Ruse 1996, 486). What more is there to be said? Sometimes, being not too bright pays off. Generations of bespectacled nerds with Mensa-level IQs have regretted bitterly that, at high school, pretty girls too often favor dumb sports players. Likewise, Conway Morris's niche suggestion draws skeptics. We have seen already in an earlier chapter that niches don't just exist, objectively, out there. They are as much created as found. What guarantee that at some point a human-like niche will necessarily turn up? Even worse, perhaps there are niches way above us where, to those niche occupants, even the nerdiest of humans seems little better than football players. Best leave things be. However, as we end this section, let us give comfort to the mechanist by pointing out that the organicist picture may be concealing a nasty hook. If we are thinking in terms of growth and development, then it is true that the Christian picture is deeply teleological. It is entirely expressed in terms of the end of time. But the teleology of the organicist model rather implies death and renewal, over and over again. This may be Buddhism. It is not Christianity. Lots of scope for dialogue here!

INTEGRATION: TEILHARD DE CHARDIN

> If this book is to be properly understood, it must be read not as a work on metaphysics, still less as a sort of theological essay, but purely and simply as a scientific treatise. The title itself indicates that. This book deals with man solely as a phenomenon; but it also deals with the whole phenomenon of man. (Teilhard de Chardin 1955, 28)

The opening lines of *The Phenomenon of Man,* by the French Jesuit paleontologist, Pierre Teilhard de Chardin. If ever they offer a prize for the most misleading opening of any book, it will be a heavy favorite. Teilhard de Chardin's book may be many things. Purely and simply, a work of science it is not. This has not stopped people from treating it as science—treating it very negatively. Thus, the Nobel Prize winner, Peter Medawar:

> The greater part of it, I shall show, is nonsense, tricked out by a variety of tedious metaphysical conceits, and its author can be excused of dishonesty only on the grounds that before deceiving others he has taken great pains to deceive himself. *The Phenomenon of Man* cannot be read without a feeling of suffocation, a gasping and flailing around for sense. (Medawar 1961, 99)

Rubbing it in:

> I have read and studied *The Phenomenon of Man* with real distress, even with despair. Instead of wringing our hands over the Human Predicament, we should attend to those parts of it which are wholly remediable, above all to the gullibility which makes it possible for people to be taken in by such a bag of tricks as this. If it were an innocent, passive gullibility it would be excusable; but all too clearly, alas, it is an active willingness to be deceived. (106)

What, then, is the *Phenomenon of Man* if not science? Theology? Not when it was published in the eyes of the Jesuits or of the Catholic Church as a whole!

> Several works of Fr. Pierre Teilhard de Chardin, some of which were posthumously published, are being edited and are gaining a good deal of success. Prescinding from a judgement about those points that concern the positive sciences, it is sufficiently clear that the above-mentioned works abound in such ambiguities and indeed even serious errors, as to offend Catholic doctrine. For this reason, the most eminent and most revered Fathers of the Holy Office exhort all Ordinaries as well as the superiors of Religious institutes, rectors of seminaries and presidents of universities, effectively to protect the minds, particularly of the youth, against the dangers presented by the works of Fr. Teilhard de Chardin and of his followers. (O'Connell 2017, referring to the Monitum [Warning] by the Holy Office, 1962)

With this kind of invective from both sides, one senses there has to be something of interest here, particularly when one learns from a recent (secular) historian that Teilhard de Chardin "was certainly the most brilliant French paleontologist of the first part of the twentieth century" (Gayon 2013, 301). On the side of religion, more recently, both the last pope and the present pope have written appreciatively of Teilhard de Chardin's thinking. In this revisionist spirit, if we criticize Teilhard de Chardin for misleadingly calling his book "science," let us praise him for offering at the time the most systematic and audacious attempt at integration—evolutionary theory and the Christian religion.

The basic thesis of *The Phenomenon of Man,* written in the 1930s although only published posthumously in 1955, is quite simple. The world and life within it are a process of evolution, of ever-greater complexity, from the inorganic through the simplest organisms up through various stages of existence to the highest, the "noösphere," the domain of humankind, culminating in something Teilhard de Chardin called the Omega Point:

> Our picture is of mankind labouring under the impulsion of an obscure instinct, so as to break out through its narrow point of emergence and submerge the earth; of thought becoming number so as to conquer all habitable space, taking precedence over all other forms of life; of mind, in other words, deploying and convoluting the layers of the noösphere. This effort at multiplication and organic expansion is, for him who can see, the summing up and final expression of human pre-history and history, from the earliest beginnings down to the present day. (Teilhard de Chardin 1955, 190)

Inventively, if very controversially, Teilhard de Chardin identified the climax, the Omega Point, with God as incarnated in Jesus Christ:

> The universe fulfilling itself in a synthesis of centres in perfect conformity with the laws of union. God, the Centre of centres. In that final vision the Christian dogma culminates. And so exactly, so perfectly does this coincide with the Omega Point that doubtless I should never have ventured to envisage the latter or formulate the hypothesis rationally if, in my consciousness as a believer, I had not found not only its speculative model but also its living reality. (293)

Inventive, but overall not entirely unfamiliar. In his Preface to the *Phenomenon of Man,* Julian Huxley writes "before being ordained priest in 1912, a reading of Bergson's *Evolution Créatrice* had helped to inspire in him a profound interest in the general facts and theories of evolution" (Teilhard de Chardin 1955, 21). The vision of evolution that is the backbone of *The Phenomenon of Man* is Bergsonian through and through—above all, organicist, in seeing ever-greater complexity, driving toward some higher end, humans! If you come at Teilhard de Chardin through the lens of a Darwinian mechanist, you may have science on your side, but you miss what he is about. Already, we know full well that his is an overall philosophical position endorsed by secular thinkers as much as by religious thinkers.

What then of the four categories or issues that have structured discussion thus far in this chapter? *The Phenomenon of Man,* written by a Christian, does not linger over Heidegger's fundamental question: "Why is there something rather than nothing?" The answer is a given. God did it. Teilhard de Chardin writes:

> To push anything back into the past is equivalent to reducing it to its simplest elements. Traced as far as possible in the direction of their origins, the last fibres of the human aggregate are lost to view and are merged in our eyes with the very stuff of the universe. (36)

At once, Teilhard de Chardin makes it clear that he is thinking in terms of the nature of the already-existing stuff, and he excuses himself of even thinking much about this because it is more a matter for physics than his own field of biology. It is all a bit complex for the nonexpert. "As I am a naturalist rather than a physicist, obviously I shall avoid dealing at length with or placing undue reliance upon these complicated and fragile edifices" (38).

Values do interest Teilhard de Chardin. In the manner of an organicist, he sees ever-greater value the closer we get to humankind, and the more humankind grows in complexity, the more value we have and the more value we want to generate:

> Our picture is of mankind labouring under the impulsion of an obscure instinct, so as to break out through its narrow point of emergence and submerge the earth; of thought becoming number so as to conquer all hab-

> itable space, taking precedence over all other forms of life; of mind, in other words, deploying and convoluting the layers of the noösphere. This effort at multiplication and organic expansion is, for him who can see, the summing up and final expression of human pre-history and history, from the earliest beginnings down to the present day. With love, as with every other sort of energy, it is within the existing datum that the lines of force must at every instant come together. (190)

Adding:

> And, conquered by the sense of the earth and human sense, hatred and internecine struggles will have disappeared in the ever-warmer radiance of Omega. Some sort of unanimity will reign over the entire mass of the noösphere. The final convergence will take place in peace. (287)

Panpsychism is the natural philosophy of consciousness for one who thinks in this mode: "we are logically forced to assume the existence in rudimentary form (in a microscopic, i.e. an infinitely diffuse, state) of some sort of psyche in every corpuscle, even in those (the mega-molecules and below) whose complexity is of such a low or modest order as to render it (the psyche) imperceptible—just as the physicist assumes and can calculate those changes of mass (utterly imperceptible to direct observation) occasioned by slow movement" (301).

And so to purpose and ends—that which gives the full (and only) meaning to Teilhard de Chardin's world system. With the noösphere, we go hand in hand with the secular organicist, and then with the Omega Point we go beyond, putting all in a Christian context:

> The end of the world: the overthrow of equilibrium, detaching the mind, fulfilled at last, from its material matrix, so that it will henceforth rest with all its weight on God-Omega. The end of the world: critical point simultaneously of emergence and emersion, of maturation and escape. (287)

All in all, a wonderful world vision, even if it must be said: *C'est magnifique, mais ce n'est pas la science.*

INTEGRATION: WHITEHEAD

> And maybe he suffers from the suffering
> Inherent to the transitory, feeling grief himself
> For the grief of shattered beaches, disembodied bones
> And claws, twisted squid, piles of ripped and tangled,
> Uprooted turtles and crowd rock crabs and Jonah crabs,
> Sand bugs, seaweed and kelp.

Thus, the American poet, Pattiann Rogers (2001, 182–183).

One thing you can say for sure. She is not writing about the God of Teilhard de Chardin. St. Augustine would find much to object to in the *Phenomenon of Man*. The idea of evolution and of progress would be out for a start. But he would recognize Teilhard de Chardin's God. He is Creator. He is the supreme endpoint. He does not Himself evolve but waits for humankind to rise to a level at which they can interact. He enters into the evolutionary process but is not part of it. No reason at all to think Him limited and developing like us. He is not one who "suffers from the suffering Inherent to the transitory, feeling grief himself." One doubts that Teilhard de Chardin's God goes quite as far as Anselm: "For when thou beholdest us in our wretchedness, we experience the effect of compassion, but thou dost not experience the feeling" (1903, 13). Or Aquinas: "To sorrow, therefore, over the misery of others does not belong to God" (1952, I, 21, 3). God, it is said, is "impassible." Quite as far as Anselm or Aquinas? Perhaps not. Nevertheless, Teilhard de Chardin's God, however much He might break from the traditional picture, is not down in the trenches, sweating out alongside us here on earth—at least, not until we get to Jesus Christ (both God and man) and the Omega Point.

It is the perceived unacceptability of this kind of God that is the raison d'être for Whitehead's approach to the God problem, developed as it was into so-called "process theology." Whitehead and his followers wanted nothing to do with a God who feels no compassion for the family when they learn that their child has leukemia. Nothing to do with the God who is unmoved—could not be moved because He is eternal and unchanging—by the death of Anne Frank in Bergen-Belsen. In any case, as an out-and-out follower of Schelling, on the one hand, Whitehead took the inherent change of organicism as all-important and, on the other hand, was totally committed to a God in the world rather than a God who is in some sense logically separate. Remember: "Nature should be Mind made visible, Mind the invisible nature. Here then, in the absolute identity of Mind in us and Nature outside us, the problem of the possibility of a Nature external to us must be resolved" (Schelling [1797] 1988, 42). Whitehead writes:

> The vicious separation of the flux from the permanence leads to the concept of an entirely static God, with eminent reality, in relation to an entirely fluent world, with deficient reality. But if the opposites, static and fluent, have once been so explained as separately to characterize diverse actualities, the interplay between the thing which is static and the things which are fluent involves contradiction at every step in its explanation. (Whitehead [1929] 1978, 346)

Continuing:

> The final summary can only be expressed in terms of a group of antitheses, whose apparent self-contradictions depend on neglect of the diverse categories of existence. In each antithesis there is a shift of meaning which converts the opposition into a contrast.
>
> It is as true to say that God is permanent and the World fluent, as that the World is permanent and God is fluent.
>
> It is as true to say that God is one and the World many, as that the World is one and God many.
>
> It is as true to say that, in comparison with the World, God is actual eminently, as that, in comparison with God, the World is actual eminently.
>
> It is as true to say that the World is immanent in God, as that God is immanent in the World.
>
> It is as true to say that God transcends the World, as that the World transcends God.
>
> It is as true to say that God creates the World, as that the World creates God. (347–348)

That means that God Himself must be part of the evolutionary process. Not the world itself—pantheism—but in all parts of the world—what

Whitehead, after Schelling, called "panentheism." And since the process of evolution involves struggle—the Englishman Whitehead is part of the Malthusian culture that produced Darwin rather than the culture of the Frenchman Teilhard de Chardin—this means that God too must be suffering and striving: "he suffers from the suffering Inherent to the transitory." He has—as the story of Jesus shows full well—emptied Himself of His powers and works alongside us. "He made himself nothing by taking the very nature of a servant, being made in human likeness" (Phil. 2:7). Kenosis.

Let us run quickly through the four topics that we argue are simply not where science gives adequate, if indeed it gives any, answers. First, origins. We do want to stress that we don't think that Teilhard de Chardin simply embraced every aspect of the traditional picture of God—obviously he didn't—but as we have seen, God as creator from nothing does seem to be part of his theological world picture. Ian Barbour, drawn to process theology, writes: "Teilhard says that God is 'self-sufficing' and initially 'stood alone.'" He denies the need for a "pre-existing substratum on which God operated and holds that matter is not eternal" (Barbour 1969, 50). Whitehead, as Barbour notes, is more radical. "Whitehead shares Teilhard's themes of continuing creation and unification, but he explicitly rejects creation out of nothing" (51). He holds that time is infinite. There was no first day, no initial act of origination, but only a continuing bringing-into-being in which past, present, and future are structurally similar. God has a priority in ontological status but no temporal priority over the world. God "is not before all creation but with all creation" (51). Elsewhere: "God is in the world or nowhere, creating continuously in us and around us. The creative principle in animate and so-called inanimate matter, in the ether, water, earth, human heart" (Price 1954, 370). Obviously, Whitehead is breaking from Plato's conception of God as the Form of the Good, eternal, unchanging. But as Whitehead himself said, humorously, all of philosophy is footnotes to Plato. One senses very much that Whitehead's God is akin to Plato's Demiurge, understood as a principle of creative design, always, on or within ever-existing matter.

As for Teilhard de Chardin, for Whitehead, ethical dicta emerge naturally from his system. We are co-creators with God. He is good, and it is for us to emulate and aid Him in His tasks. Since everything is part of the whole creative system, interrelated in a whole, our obligations stem naturally from this. The leading process theologian, John Cobb, writes that "a human life is 'of more value than many sparrows' (Matthew 10:31) does not warrant the conclusion that sparrows are worth nothing at all. Indeed, it presupposes the opposite. The Heavenly Father cares even for sparrows; how much more for human beings! This certainly means that people too should be concerned more about a human being than a sparrow. Much more! But it does not warrant the teaching that sparrows exist only as a means to human ends . . . God is pictured as loving the creatures and caring for them, not only human beings, but sparrows as well" (Cobb 1991, 44). To be honest, one does sometimes get the feeling that process theologians rather make up their moral imperatives—"substantive ethics"—as they go along. Expectedly, this ethics is much in line with liberal thinking today—tolerance of those who are gay and the like. One can see how and why a process theologian might approve of killing mosquitoes. After all, they do threaten human beings, and we are more important than they. But, to return to a worry expressed in Chapter 2, if not a vegan, it is not obvious that one can refrain from being a vegetarian. Is sitting down to a rare T-bone steak "loving the creatures and caring for them"? A cattle farmer, of course, would say that beef cattle are cared for.

The problem of consciousness has been covered already. Being a panpsychist comes with the territory—which brings us to the interesting matter of purpose and ends. For Teilhard de Chardin, the story is one of increasing complexification, going up through that noösphere and ending with the Omega Point, God made flesh in Jesus Christ. This is not exactly predestined in a Calvinistic sort of way, but it is the way things are going to end. And—Plato again—this is going to be an all-inclusive unity. Ian Barbour writes: "Teilhard sees the whole cosmic process as one slowly culminating event with a single goal." Continuing, explicating *The unity of all things in Christ:*

> Teilhard's idea of the "cosmic Christ" combines his conviction of the organic interdependence of the world and his biblical belief in the centrality of Christ. Redemption is not the rescue of individuals from the world but the fulfilment of the world's potentialities; the corporate salvation of the cosmos is integral with the activity of continuing creation. The world converges to a spiritual union with God in Christ, whose relation to the world is organic and not merely juridical and extrinsic. (Barbour 1969, 57)

It is worth emphasizing, since so often Teilhard de Chardin is taken as heretically throwing away the very essence of Christian doctrine, how very orthodox in this matter Teilhard de Chardin really is. Writing around the same time, his fellow Jesuit, the theologian Karl Rahner, took up the question of evolution and Christianity. We learn that humans are the "self-transcendence of living matter," meaning we are thinking beings aware of ourselves and our situation, and that "the history of Nature and spirit forms an inner, graded unity in which natural history develops towards man, continues in him as his history, is conserved and surpassed in him and hence reaches its proper goal with and in the history of the human spirit" (King 1996, 8). We are the only truly free beings. This doesn't happen by chance:

> In so far as this history of freedom, however, always remains based on the pre-determined structure of the living world, and in so far as (as the Christian professes) the freedom-history of the spirit is enveloped by the grace of God which perseveres victoriously unto the good, the Christian knows that this history of the cosmos as a whole will find its real consummation despite, in and through the freedom of man, and that its finality as a whole will also be its consummation. (Rahner 1966, 168)

Apart from affirming Teilhard de Chardin's orthodoxy in this respect, what is revealing is that, in a forty-page essay on the topic, Rahner seems not to have felt the need to read a word of professional scientific thinking on the evolution question. One feels that, in any case, it would not have been necessary, for all Rahner is about to do is simply whip evolution into line. Conservative in respects or not, Teilhard de Chardin is going the other way, trying to see how Christianity can be rethought in the light of evolution. As, of course, was precisely the intent of Whitehead who, unconstrained by a Jesuit education, felt free to break with this kind of thinking entirely. We are co-creating with God, and that means there can be no absolute guarantee of success:

> Maybe he wakes periodically at night,
> Wiping away the tears he doesn't know
> He has cried in his sleep, not having had time yet to tell

> Himself precisely how it is he must mourn, not having had time yet
> To elicit from his creation its invention
> Of his own solace. (Rogers 2001, 182–183)

Process pastor and theologian Bruce G. Epperly (2011) grasps the nettle, as one might say in an appropriate metaphor. We humans could let off atom bombs, Dr. Strangelove style, and end it all. Not to worry!

> Humans are loved by God who seeks creatures of stature and complexity; but God also loves the whole universe and seeks beauty in every planetary and galactic context. If humans choose the pathway of self-destruction, God will nurture other streams of evolution on earth and other planets. (99)

We have only ourselves to blame. And whether there is life after death, to make up for it all, is not a done deal. "At present it is generally held that a purely spiritual being is necessarily immortal. The doctrine here developed gives no warrant for such belief. It is entirely neutral on the question of immortality" (Epperly 2011, 136–137, quoting Whitehead 1960, 107).

On that somber note, we bring this chapter to an end. We predicted that, for the religious, the organic metaphor would prove more attractive than the machine metaphor. We have shown that this is true. Most who approach or interpret life from a religious perspective find the organic metaphor far more congenial—which is no great surprise, for it is that metaphor that incorporates value and that makes human beings all important. No surprise either that, from the start, Darwin's theory has generally been considered unfriendly to religion, especially Christianity—although note the nature of this unfriendliness. Not all thought Darwin's theory refuted any possible belief. In writing the *Origin,* we saw that Darwin himself was a deist. But it was taken as corrosive of belief. God may still exist, but He now seems to have become less relevant. Thomas Hardy, raised a good Anglican, raised just these doubts. This sonnet "Hap" was written in 1866:

> If but some vengeful god would call to me
> From up the sky, and laugh: "Thou suffering thing,
> Know that thy sorrow is my ecstasy,

> That thy love's loss is my hate's profiting!"
> Then would I bear it, clench myself, and die,
> Steeled by the sense of ire unmerited;
> Half-eased in that a Powerfuller than I
> Had willed and meted me the tears I shed.
>
> But not so. How arrives it joy lies slain,
> And why unblooms the best hope ever sown?
> —Crass Casualty obstructs the sun and rain,
> And dicing Time for gladness casts a moan. . . .
> These purblind Doomsters had as readily strown
> Blisses about my pilgrimage as pain. (Hardy 1994, 5)

You might now conclude that the person of faith must embrace the New Biology and reject Darwinism and mechanism generally. Or that they must play the trick of being a methodological naturalist from Monday to Friday, and a metaphysical nonnaturalist at the weekend. But before you rush to final judgment, although perhaps a minority position, be aware that all would not follow you. Remember Aubrey Moore: "Darwinism appeared, and, under the guise of a foe, did the work of a friend" (1890, 99). This is hardly yet proof that there is still life in the Darwinism-Christianity (or any other religion) exchange. At best, we regard Moore's optimism as a program and a challenge, not a complete answer, although it is an optimism that many contemporary science-theologians share, and not just within the Christian tradition (Southgate 2008).

Do not conclude that, to the contrary, we are pessimistic. As always in this book, our aim is to give you the tools to see the possibilities and perhaps make a decision. In this chapter, we have seen times where organicism has the same problems as mechanism, and perhaps indeed mechanism escapes some of the problems of organicism. This quite apart from the question of which leads to better and more fruitful science. Whatever empathy we show in this book to the New Biology, the vast majority of practicing professional evolutionists are hardline Darwinians, well satisfied with their paradigm. Speaking against the organicist philosopher John Dupré, quoted earlier in this chapter: "As an evolutionary biologist—which Dupré is not—I think I'd know if my field was in crisis. Yet I haven't heard any recent lamentations from my colleagues" (Coyne 2012). We do say that, if supporters of the machine metaphor, Darwinians, want to go on being Christians—or members of other faiths—or if Christians want to be in line with what most professionals would think the best-quality science—then, unlike Karl Rahner, they had better start taking the challenge of evolutionary thinking as seriously as did Pierre Teilhard de Chardin and Alfred North Whitehead.

Chapter Nine

BIOLOGY EDUCATION

> This idea of purity and you're never compromised and you're politically woke, and all that stuff—you should get over that quickly. The world is messy. There are ambiguities. People who do really good stuff have flaws.
>
> —BARACK OBAMA 2019

As we hope is clear from the chapters to date, the two of us are passionate about biology as a discipline and about its relevance to humanity. We see the advances made by biology as key to helping all of us better understand ourselves as a species, and vital if humanity is to get through the twenty-first century without destroying itself or even more of the rest of nature and the environment. But we are concerned that the way that biology is taught and communicated, in schools and elsewhere, too often presents only a partial account of the discipline. Often, a narrow, somewhat reductionist, and overly mechanistic account of the subject is favored. As we have argued, it is not that we reject the usefulness of reductionist and mechanistic approaches as tools to understanding life. Rather, our objection is that on their own, these approaches are only part of what is needed to understand biology. Equally, an approach that focuses only at too high a level can be fluffy and fail to make falsifiable predictions. We believe both are needed. Biological issues need to be examined at a number of levels—at the behavior of whole organisms and the elements of which they are composed (down to the molecular level), and at aggregates of organisms (into populations, ecosystems, and beyond), as well as at the mechanisms that enable these various levels to function.

As we have shown, there are deep divisions among biologists—one only has to look at the controversy over the unit of selection in the field of evolutionary biology and population genetics that we discussed in Chapter 2. The way that some former comrades-in-arms now speak and write about each other's work reminds one of the propaganda purges and obliteration from history (albeit not the murders) that characterized the peak of Stalinism. Interestingly, very little of this intellectual controversy filters down to school level, which all too often presents a sanitized account of the discipline—one that gives the impression that biologists are in agreement about their subject and that the story is one of near uninterrupted progress as biology constantly advances. No wonder many intelligent youngsters leave science when they can, preferring subjects with a diversity of voices. This is in marked contrast to some sections of the media that, favoring an adversarial approach to issues, sometimes overstate disagreements between scientists, which can give the impression that almost no biological knowledge is robust and trustworthy.

Biological knowledge does advance, but the lessons from history are that it's not a simple story of "onward and upward." Furthermore, as we have argued, much of the disagreement between the various camps cannot be resolved simply though the day-to-day workings of science—there are fundamental and incommensurable differences in terms of how people see issues. Different groups adhere to different metaphorical ways of understanding reality. Disagreements between biologists are not just temporary ones, existing for a short passage of time until new data are collected that unambiguously resolve the dispute. There are instances where that is the case, but there are also many instances where the disagreements are more deep-seated, to do with how biologists, indeed each of us, understands the world; people occupy different worldviews.

We saw this, for example, in our chapter on sex and gender. If you see gender as proceeding unproblematically from sex, then, setting aside the quite small number of people who are intersex, the overwhelming majority of humans do conveniently and unambiguously divide into males and females, and the whole transgender debate makes little sense.

However, if you see gender as more to do with identity and how one sees oneself, then a universal binary classification into males and females doesn't work. The difference between these two positions won't be resolved by peering at chromosomes down a microscope or taking measurements of testosterone and estrogen levels.

This chapter is therefore addressed to a diversity of audiences. Chiefly, we are writing to those who control what is taught in education systems—in schools especially but also in further and higher education and in museums and other sources of out-of-school learning. What is taught in education systems is itself the result of interactions at a range of levels within what we might term "the education ecosystem." Think of a student in a class in school. The most important influences on what that student learns are the student and the teacher(s) (e.g., Hattie et al. 2020). Other influences on a student's learning are home background, other students in the class, the school itself (beyond its teachers—things such as the resources it has and the curricula it uses), and various factors beyond the school level (such as educational policies at local or national level). Furthermore, these levels are codependent; by and large, schools with better facilities and a history of success are more likely to attract teachers who are seen as good teachers. If parents have some degree of choice as to where their children are educated, this makes such schools more popular to many parents, and one enters something of a positive feedback loop in which it becomes doubly difficult for less successful schools to "catch up."

To illustrate how the debate between reductionism and holism plays out in real life, we will therefore explore biology education, both school biology education and out-of-school biology education for learners of all ages.

SCHOOL LABORATORIES

Imagine a school laboratory used for teaching biology. As a school biology teacher, you want your students to learn about the growth of plants. This is a topic in school biology courses around the world, and photosynthesis plays a large part in what is taught. We won't launch into a full, school-level account of the activities a biology teacher might get their students to undertake, but chances are, when teaching about plant growth, you would want your students to understand the following:

- Leaf structures are appropriate for their functions. So, most leaves have tiny holes in their surfaces that allow gases to enter and leave. During the day, these holes (stomata) are open because carbon dioxide is needed for photosynthesis, but at night, or if it gets very dry, these holes shut, thereby reducing the evaporation of water from the leaves.
- The carbon dioxide that has entered the leaves reacts with water in the chloroplasts in the presence of the green pigment chlorophyll. This is a chemical reaction that requires a lot of energy, provided by light (whether sunlight or artificial light). The two main products of photosynthesis are sugars and oxygen. To the plant, the oxygen is pretty much a waste product (though some oxygen is used in other chemical reactions), but the sugars are vital. They are combined with one another to make large carbohydrates, such as starch, which are stored (think potato tubers), or with nitrogen and other minerals obtained from the soil via root hairs to make amino acids (the building blocks of proteins) and other molecules needed by the plant.

In a school laboratory, quite a bit of this can be taught using simple equipment. You can get students to look at leaf structures through light microscopes, you can get them to undertake experiments to show that plants take up carbon dioxide from the atmosphere and produce carbohydrates (such as sugars and starch) and oxygen, and you can demonstrate that light and chlorophyll play essential roles in photosynthesis.

However, the laboratory provides a simplified, stripped-down model of reality; it provides an environment in which reality is "tidied up," laid bare, reduced to its essentials so that certain scientific phenomena can be studied much more readily than they ever could in "the real world." It is more difficult in an actual woodland than in a laboratory to show that plants take in carbon dioxide and produce carbohydrates and oxygen.

Second, many school students completely fail to understand why they are undertaking experiments to do with potassium hydroxide (which absorbs carbon dioxide) or iodine solution (which tests for starch). The reductionist approach in which the process of plant growth is divided up into a set of stages and each one explored may make complete sense to a curriculum developer and to biology teachers but too often results in students not learning what their teachers want them to learn (Sanders & Jenkins 2018). If one takes a clock to pieces to show how it works, at some point, one needs to put it back together again—and that's not an easy task.

Furthermore, there are more ways to study biological systems than there are ways to study clocks. Consider a forest. A biologist would be most interested in the organisms in the forest, a climatologist would study such things as rainfall, sunlight, topography, and wind, and a geologist would focus on the underlying rocks and the consequences of these for the soil (Reiss & Tunnicliffe 2001). Even if we put to one side such obvious species-specific roles occupied by those who define themselves as microbiologists, botanists, mycologists, and zoologists, our forest might be packed with ecologists, anatomists, biochemists, physiologists, and those interested in the history of the forest as revealed by a variety of different approaches including tree-ring dating, field archaeology, and the study of place names (Rackham 2015).

Indeed, we can subdivide further: our ecologists will include population biologists (spending their time determining the numbers of individuals within species and organizing these individuals by age classes), ecological geneticists (concerned with any relationships between individuals' genetic make-up and their fitness), autecologists (each concentrating on the ecology of a single species), synecologists (attempting to unravel the interrelationships between species and therefore looking at the biology of the forest as a whole), conservation biologists (concerned to prevent, through careful management based on thorough monitoring, the loss of rare species from the forest), and so on and so on. Each of these specialists has a partial understanding of the forest. We need a biology education that helps learners to see the wood and the trees.

We will now look at two issues: vaccine hesitancy, and nutrition education. In each case, we will discuss how the issue benefits from a multilevel analysis that makes use of a more holistic approach than is often the case, whether we are talking about school education or out-of-school education, such as that provided in the media or in science museums.

VACCINE HESITANCY

At the time of writing, vaccines against COVID-19 are being rolled out for the public in an increasing number of countries in a battle against the virus itself and its latest variants. Yet, vaccines are not universally welcomed—and this is not only the case for vaccines against COVID-19. Vaccine hesitancy or rejection occurs despite the fact that vaccines are one of medicine's great success stories, preventing (pre-COVID) an estimated two to three million deaths a year (World Health Organization 2019b). However, vaccination rates were falling even before COVID-19 arrived on the scene. There are a number of reasons for this fall. Some of it resulted from the now refuted suggestion that autism rates were rising as a result of vaccination against measles, mumps, and rubella (MMR). In the Philippines and Indonesia, vaccine confidence plummeted in 2017 when the vaccine manufacturer Sanofi announced that its newly introduced dengue vaccine posed a risk to individuals who had not previously been exposed to the virus. In South Korea, an online community named ANAKI (Korean abbreviation of "raising children without medication") strongly argued against childhood immunization at this time, which dented public confidence. In 2019 (i.e., pre-COVID), the World Health Organization identified vaccine hesitancy as one of the ten threats to global health, with measles, for example, seeing a 30 percent increase in cases (World Health Organization 2019b).

Our argument is that an approach to vaccine education that takes seriously the objections that some have to their deployment would be better education. Too often, an implicitly reductionist approach is taken to vaccine education—one that seems to think that it's enough to demonstrate the benefits of vaccines for the body's immune system. As a result, vaccines are usually presented in school biology lessons, medical establishments, and government pronouncements as an unqualified good, so that those who object to their use are branded ignorant or selfish. This is not good education—for one thing, such practices are more likely to lead to entrenched attitudes than to help people change their minds and behaviors.

Vaccination (strictly, immunization unless one is talking specifically about smallpox) has benefited the lives of huge numbers of humans and nonhuman animals (Greenwood 2014) for more than a century. The one human illness that immunization has succeeded in eliminating is smallpox. Smallpox may have originated some 3,000–4,000 years ago in East Africa, its origins possibly triggered by the introduction of camels and concurrent changes to the climate that led to its evolution from a cowpox-like ancestral virus (Babkin & Babkina 2015). The advent of smallpox had major consequences for individuals and for societies. Its typical mortality rate was around 20 percent, although for infants, this figure was nearer 80 percent and, in some circumstances (e.g., Berlin in the 1800s), reached 98 percent for infants (Riedel 2005). Smallpox played a part in the decline of the Roman Empire—the Antonine Plague (from 165 CE) killed almost seven million people—and in the fall of the Aztec and Inca civilizations, having been introduced to South America by the Spanish and Portuguese conquistadors. In eighteenth-century Europe, 400,000 people died annually from it, and one-third of survivors went blind.

As early as 430 BCE, survivors of smallpox were called upon to nurse the afflicted; it was realized that if you had caught smallpox (and survived), you were unlikely to catch it a second time. What became known as variolation was practiced in Africa, India, and China long before it was introduced to Europe. In the fifteenth century, the practice of nasal insufflation occurred in China, in which powdered smallpox material was blown into one of the nostrils—the left nostril for girls and the right for boys (Williams 2010). People who had recovered from mild cases of the disease were used as donors, and scabs that had been left for some time were used, since fresh scabs were known to be more likely to lead to fatalities.

Variolation had a fatality rate of about 2–3 percent (Riedel 2005), almost an order of magnitude less than smallpox but not insignificant. The next leap forward came, as is well known, because the doctor Edward Jenner took seriously the folktale that dairy maids owed their traditional fine complexions to the fact that once they had caught the relatively minor disease cowpox, as many of them did, they were immune from smallpox. In 1796, Jenner inoculated eight-year-old James Phipps with pus from the hand of Sarah Nelms, a milkmaid who had contracted cowpox from a cow called Blossom—you can see one of Blossom's horns at Dr. Jenner's House, Museum and Garden in Berkeley, Gloucestershire. Phipps caught cowpox, and after he had recovered, Jenner exposed him to smallpox—medical ethics operated somewhat differently then. Thankfully, Phipps did not develop smallpox, and as a consequence of Jenner's work, vaccination—a term that Jenner devised from the Latin for cow, *vacca*—eventually led to the worldwide eradication of smallpox in 1977.

Objections to vaccination began almost as soon as the practice was introduced. Nineteenth-century objections included the arguments that vaccination did not work or was unsafe (Ernst & Jacobs 2012) and that its compulsory introduction (e.g., the 1853 Compulsory Vaccination Act in the UK) violated personal liberties (Durbach 2000). Today's objections to vaccination overlap with these and include the following:

- Vaccines don't always work
- Vaccines are not totally safe
- Requiring, or even just incentivizing, someone to be vaccinated or to have their child(ren) vaccinated violates personal liberties
- Vaccines are often made in ways that are morally unacceptable
- The scientists or companies that make vaccines can't be trusted
- Governments that advocate vaccination uptake can't be trusted
- Vaccines are unnatural
- Vaccines are part of a conspiracy to poison us, take over our minds, or control us in some other way.

What is notable about these objections to vaccination is how limited the role of science is in addressing them. Consider the first two—that vaccines don't always work and that they are not totally safe. The objectors are not saying that a cost-benefit analysis on the grounds of efficacy or safety comes down against the use of vaccines—indeed, such cost-benefit analyses very strong support vaccine efficacy and safety. Rather, the objectors are saying that vaccines don't *always* work and aren't *totally* safe. You can't argue against these objections on scientific grounds; even the most reliable of medical interventions sometimes do not work, and sometimes they produce harmful effects.

Other objections have even less to do with science. Consider the objection that vaccines are often made in ways that are morally unacceptable. It all comes down to what one means by "morally acceptable." For example, a number of widely used vaccines use cells lines derived from fetuses that were electively aborted decades ago. Examples of such vaccines are ones used against rubella (German measles), hepatitis, polio, and chickenpox (Pelčić et al. 2016). The moral debates about the use of cell lines from fetuses mirror those about abortion. These differences of opinion—deeply held convictions—cannot be reconciled by any method of science. They simply lie outside of science, situated in the domain of moral philosophy, or values more generally. This is not to say that biology education, inside or outside the classroom, does not have a role to play in the specific issue of debates about vaccinations. Physiology has a lot to say about whether fetuses are capable of experiencing pain, and psychology has something to say about the long-term consequences for pregnant women of having or not having abortions.

In fact, there might be much to be said for psychology being taught as part of or alongside biology in schools. Take conspiracy theories against vaccination. To most people, these sound so bizarre that they just get dismissed. But many people believe them. A May 2020 YouGov poll of 1,640 people found that 28 percent of US citizens said they believed that Bill Gates wants to use vaccines to implant microchips in people—with the figure rising to 44 percent among Republicans (Sanders 2020). The source of the story may be a study, funded by The Gates Foundation, into a technology that could store someone's vaccine records in a special ink administered at the same time as an injection (Goodman & Carmichael 2020).

A common objection to vaccination is that it is "unnatural." The argument about whether something is natural is one we have considered previously in our discussion of Golden Rice in Chapter 7. In one sense, vaccination is indeed unnatural—and many parents find it difficult to be with their child while another person sticks a needle into them, injecting them with something that contains parts of a disease-causing organism. And yet isn't all medicine unnatural? Or is the natural state of affairs for humans to use our abilities (some would write our God-given abilities) to do what we can to ameliorate suffering and enable people to flourish? Again, these are not questions that science can definitively answer.

Existing approaches to vaccine education seem to be designed with the intention that they will increase vaccine uptake (e.g., Edwards et al. 2016). Present methods of vaccine education typically consist of providing parents who are reluctant to vaccinate their children with information intended to get them to change their minds. Such approaches generally do not work (Navin et al. 2019). In one carefully undertaken study designed to test the effectiveness of messages intended to reduce vaccine misperceptions and increase vaccination rates for MMR, a nationally representative experiment was conducted with 1,759 parents residing in the United States who had children in their homes aged seventeen years or younger (Nyhan et al. 2014). Parents were randomly assigned to one of five groups, receiving: (i) information from the Centers for Disease Control and Prevention explaining the lack of evidence that MMR causes autism; (ii) textual information from the Vaccine Information Statement about the dangers of the diseases prevented by MMR vaccination; (iii) images of children who have diseases that can be prevented by the MMR vaccine; (iv) a dramatic narrative about an infant who almost died of measles from a Centers for Disease Control and Prevention fact sheet; or (v) none of these (i.e., a control group). As the authors conclude: "None of the interventions increased parental intent to vaccinate a future child" (Nyhan et al. 2014, e835). So, what can be done to provide better education about vaccinations?

We have three suggestions. First, we believe the aim should be to enable learners to understand more fully the consequences, for themselves and for others, of vaccine uptake and non-uptake—in other words, increase understanding. Of course, we would both be delighted if vaccine education led to greater vaccine uptake, but there is an important issue here that, while it lies within the field of moral philosophy more than biology, has certain parallels with the tension between mechanism and organicism that we have been exploring in biology. On the one hand, we both have a high opinion of people's autonomy, and if someone, even though well informed about the consequences of being vaccinated or not, decides against vaccination, it could be argued that that is their prerogative. On the other hand, it's one thing for me to decide not to get myself vaccinated against COVID-19, measles, polio, or whatever, but do I have the right to prevent my children from getting vaccinated against such dangerous diseases, given that medical opinion is overwhelmingly on the side of vaccination? A good vaccine education would help learners to consider such arguments.

Our second suggestion as to how vaccine education might be improved, one that follows from the first, is that, far more than the provision of information, vaccine education should allow learners to discuss, explore, and develop their views. We realize that this is time-consuming and therefore more difficult to arrange with adults than with students in full-time schooling, but few adults change their practices simply because they are provided with information that doesn't agree with their existing views. What we need are fora where people with opposing views about vaccination do not simply shout at one another and denigrate each other's views but rather listen to each other and discuss. There is actually quite a lot of agreement between opposing camps. Just about everyone wants to be healthier; the disagreement is largely about whether vaccines help or hinder in this.

Finally, vaccine educators should be sensitive to those who do not want vaccinations to take place. Not only is this respectful of people, but it may also be more likely to increase vaccine uptake than haranguing people is (cf. Reiss 2019). Parents who choose not to vaccinate their children almost always have their children's best interests at heart. They are unlikely to change their minds or their practices if they are faced with educators who clearly believe that they are wrong or stupid and are endangering their children's lives. Of course, just because a parent has their children's best interests *at heart* does not necessarily mean that they are right. Parents may sincerely believe they are doing the right thing to provide their daughters with less education than their sons, or to bring their children up to believe that communists are beyond the pale—or capitalists, or religious believers, or atheists, or foreigners, or whoever. Some countries have the "welfare of the child" as their guiding principle when it comes to the courts and social services deciding what should be done in difficult circumstances, such as making decisions about custody arrangements in contested divorces or deciding whether to leave potentially at-risk children within their families or take them into care. The courts and social services don't always get particular decisions right, but the overall intention is there.

NUTRITION EDUCATION

Humans have an impressive ability to consume a very wide range of organic matter and build ourselves out of it (Reiss 2020). Until we began to develop agriculture some 12,000 years ago, we got our food, as a small number of hunter-gatherers still do, by gathering plant matter and hunting animals—mainly mammals, birds, and fish, although some communities also ate other foods, such as oysters. Over the past few decades, there have been a spate of books advocating "paleo diets." The popularity of such approaches stems from a growing dissatisfaction for many people with what they are eating. However, there are problems with the paleo-diet approach. For one thing, it is quite difficult to get a diet that is close to what our ancestors were eating at this time (the Paleolithic lasted from about three million to about 12,000 years ago). Domesticated animals (bred since the end of the Paleolithic) have a lot more fat than wild ones do, while domesticated plants (also bred since the end of the Paleolithic) are a lot more digestible than are most wild plants. Certain foods that are a standard component of most people's diets nowadays (e.g., cereals and legumes) barely featured then. The fruits were smaller and tarter, and potatoes (restricted to South America) were the size of today's peanuts. A paleo diet contains little or no alcohol, tea, coffee, and chocolate. Furthermore, the ancestors of some of our domestic plants can be quite dangerous—for instance, wild almonds produce dangerously high levels of cyanide.

Most of us have diets nowadays that are higher in salt and sugars and lower in fiber than did our Paleolithic ancestors. So, paleo diets do have certain features to commend them. As is well known, many people have a greater liking for saturated fats and for sugars than is healthy. This can cause problems for those of us who now live with a superabundance of foods rich in such ingredients. In the past, when these foods were rarer, there was little to be gained—and much to be risked—in *not* stuffing ourselves when an occasional glut of such foods was found.

What Do We Mean by "A Good Diet"?

The simplest answer to the question of what is meant by good when we speak of a good diet is that such a diet is good for our physical health. Now, there is much to be said in favor of good physical health. But to focus only on physical health is a rather narrow way to envisage food. For one thing, eating is often a social activity. If we eat rather more than might strictly be good for us (in terms of physical health) or even consume more alcohol (within reason) than is medically ideal when in the company of friends or family at an occasional celebratory event, such as a birthday, wedding, or when your authors finally hand this manuscript over to the publisher, are we unwise to have done so? So, one way of understanding the phrase "a good diet" is to see it as meaning a diet that helps us feel positive about ourselves, a diet that promotes our well-being (including but more than physical health).

Even if we restrict ourselves to physical health, what constitutes a good diet remains surprisingly unclear once we get beyond the advice to eat a balanced diet, without losing or gaining too much weight. Dietary advice about fats and carbohydrates has changed over the last few decades. A recent review of the nutritional benefits and harms of different sorts of fats and carbohydrates, undertaken by a most reputable group of academics and published in a prestigious academic journal, identifies a number of relatively simple questions to which we still lack answers:

- Do high-fat, low-carbohydrate (ketogenic) diets provide health benefits beyond those of moderate carbohydrate restriction?
- What are the optimal amounts of specific fatty acids (saturated, monounsaturated, polyunsaturated) in the context of a very-low-carbohydrate diet?
- What is the relative importance for cardiovascular and other diseases of the amounts of LDL (low-density lipoprotein) cholesterol (so-called "bad" cholesterol) and HDL (high-density lipoprotein) cholesterol (so-called "good" cholesterol) in the blood?

(Ludwig et al. 2018)

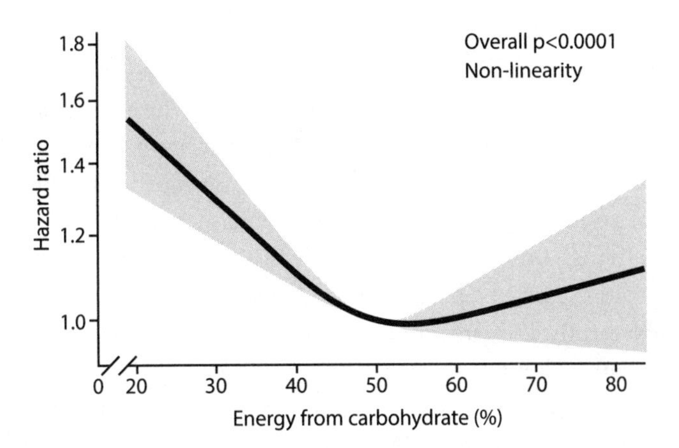

FIGURE 9.1 The U-shaped relationship between the proportion of the energy in the diet that comes from carbohydrates and mortality rates adjusted for a wide range of variables, including age, sex, ethnicity, total energy consumption, diabetes, cigarette smoking, physical activity, income level, and education. *Credit:* Reformatted from Sara B. Seidelmann, Brian Claggett, Susan Cheng, Mir Henglin, Amil Shah, Lyn M. Steffen, Aaron R. Folsom, Eric B. Rimm, Walter C. Willett, and Scott D. Solomon, "Dietary carbohydrate intake and mortality: A prospective cohort study and meta-analysis," *Lancet Public Health,* 3: e419–e428, 2018, Figure 1. CC BY-NC-ND 4.0.

As a public service announcement, we should emphasize that there is plenty of nutritional advice about carbohydrates and fats in our diet that is well evidenced and where a scientific consensus does exist, even if the scientific consensus is not widely known. For example, for those of you in favor of ketogenic or high-carbohydrate diets, did you know that diets that are either very low or very high in carbohydrates are associated with greater mortality than ones with intermediate amounts (Seidelmann et al. 2018)? Figure 9.1 shows the findings of a twenty-five-year US study: there is a U-shaped relationship between the proportion of the energy in people's diet that comes from carbohydrates and their mortality rates adjusted for a wide range of variables, including age, sex, ethnicity, total energy consumption, diabetes, cigarette smoking, physical activity, income level, and education.

But there is more to a good diet than survival or even enjoying a diet that promotes our well-being. If we focus on school education about nutrition and diet, we might also want students to think about the animal welfare implications of consuming meat from domesticated animals.

Literally tens of billions of chickens, cows, sheep, pigs, and rabbits are raised and slaughtered for food across the world each year. A small proportion of these lead good lives in terms of their welfare, but the majority don't. The UK Government Advisory Body, the Farm Animal Welfare Council, on which one of us sat as the ethicist for nine years, argued that the absolute minimum acceptable standard was for an animal to have "a life worth living" on the grounds that if this standard was not met, the animal would literally be better off dead (Farm Animal Welfare Council 2009). Regretfully, in many countries there are many domesticated animals whose lives don't even reach that standard.

By and large, the more that the circumstances in which farm animals are kept depart from nature, the worse their welfare. There are important exceptions—farm animals are protected from predators and often receive treatment for parasites and diseases. However, the high population densities in which they are usually kept bring new disease problems and lead to raised stress levels and an inability to manifest many of their natural behaviors. Anyone who equates nature with the good is likely to consider all but the best of farm animal treatment unacceptable. When one considers the problems caused by the domestication of farm animals, two major responses can be identified. One is that we should go back to a time when farmed animals were treated more naturally—kept at lower population densities, subject to normal lighting regimes, allowed to go outside, able to exhibit a full repertoire of behaviors, and so on. The other is that each of the problems can be tackled by further interventions. So, if high population densities cause greater rates of disease transmission, prophylactically give all the animals disease-preventing treatments; if high population densities cause chickens painfully to peck one another, debeak chickens; if farming practices cause animals to experience high levels of stress, breed them so that they are more docile when overcrowded or pushed around; and so on.

What Might Good Education about Nutrition Entail?

Good education about nutrition entails learning and thinking about food at a number of levels and from a number of perspectives. If we start with what is conventionally taught in schools, the cornerstone of nutrition education is an understanding of "a balanced diet." However, a balanced diet, while more affordable for most people nowadays than in the past, can still be expensive. In the UK, a healthy balanced diet for the poorer half of the population would account, on average, for close to 30 percent of disposable income; this compares with an average of 12 percent of disposable income for the wealthier half of households (Scott et al. 2018). Internationally, food inequalities are even greater.

Talk of such inequalities suggests that we might want a good nutrition education to include consideration of ethical issues—something to which we have already alluded in our consideration of farm animal welfare issues. Nor should it be supposed that it is straightforward to address ethical issues. The moral framework within which contentious issues are examined makes a difference—one will reach different conclusions about both human inequalities and farm animal welfare depending, for example, whether one holds a utilitarian or a right-based position.

Morality is close to values, and school biology education generally doesn't tend to do a particularly good job at enabling students to reflect on and develop their values. And yet values are particularly closely intertwined with the decisions we make about what we do or don't eat. Even setting aside religious prohibitions on diet, students may be vegetarians for a range of reasons: because they believe it's wrong to eat meat, because they believe that a vegetarian or vegan diet is healthier, because they are concerned about sustainability issues (soil erosion, climate change, conversion of natural vegetation to land for animal farming), because they believe it will help tackle world hunger, because they don't like the appearance or taste of meat, and so on. Some of these reasons are only tangentially related to the main themes of this book, others centrally so. It would be good if school nutrition education could allow them and others to be considered.

GOOD BIOLOGY EDUCATION

In attempting to explain behaviors in humans, we need both bottom-up and top-down explanations to understand what is going on. Consider one of the cases we have examined in this chapter: nutrition education. It is often said that "we are what we eat," and there is an element of truth in this in that a mechanistic approach to our diet can, to a certain extent, predict the consequences for us of eating too much or too little overall or of having an unbalanced diet. One of your two authors managed to develop scurvy as a child as a result of insufficient vitamin C in his diet one winter, which was pretty embarrassing for his parents, one of whom was a doctor and the other of whom had been a nurse and midwife. Treating the body as a machine that lacks a vital supply, the scurvy cleared up within a week with healthy doses of vitamin C.

But while our diet has consequences for who we are, who we are has consequences for our diet. We need both bottom-up and top-down explanations to understand what we eat and the consequences of this for us. Anorexia nervosa is not adequately explained as a diet deficient in energy—we want to know why a person takes in insufficient calories even when there are plenty available. Similarly, in relation to another of our cases, vaccines have major, potentially lifesaving, consequences for who we are, *and* who we are helps determine whether we agree to be vaccinated or allow our children to be vaccinated. In biology, perhaps more than in any other science, our values and our worldviews play core roles in how we understand the subject. Good education can help learners appreciate this and see other perspectives.

A good school biology education should help students appreciate the value of both the mechanism metaphor and the organism metaphor, and the benefits of sometimes taking a systems approach (Gilissen et al. 2020). Again, nutrition education illustrates this well. A mechanistic approach allowed biochemists and other scientists to realize that rickets normally results from a shortage of vitamin D or calcium. In the 1970s, it became clear in Britain that rickets was making a comeback in certain cities, including Bradford, Birmingham, and Glasgow. These cities had large and expanding ethnic minority populations, and studies were undertaken to establish why rickets was more likely in these groups than in the majority population (Bivins 2007). Because vitamin D can be obtained both in the diet and made by the action of sunlight, understanding *who* develops rickets in Britain benefits from a whole organism approach. Such an approach takes seriously differences between people with respect to their diet (e.g., whether they are vegetarian—vitamin D being found principally in oily fish, red meat, and egg yolks), their clothing, and their behavior—with all of these often depending on ethnicity, gender, and culture.

A good biology education, one that considers the issues we have been addressing, demands more, intellectually, from learners than that which biology education often expects. It also gives learners more agency, as it's not a matter of learners simply being presented with information about genetics, nutrition, ecosystems, or whatever, but rather of helping them to see what is going on at the various levels and to appreciate the interactions that are happening within levels too.

A rich biology education can therefore help students in schools, and learners in other situations, to develop a fuller understanding of biological issues, particularly when these are so-called socio-scientific issues where understanding requires a good grasp not only of the biology (understood narrowly) but also of the social context within which the biology is situated. In our examples above, vaccine hesitancy and nutrition education are both examples of socio-scientific issues. So too are such topical issues as climate change, loss of biodiversity, and biotechnologies, education about all of which, we would argue, would benefit from the approach we have been advocating.

EPILOGUE

We have titled this book *The New Biology: A Battle Between Mechanism and Organicism*. Our fundamental argument begins with the assertion that both mechanism and reductionism, on the one hand, and organicism and holism, on the other, are different ways of looking at and attempting to understand biological systems. At different times, one or other of these viewpoints has held sway. Our point is that on their own, each is generally insufficient if all aspects of the system are to be understood. Often, for example, when teaching biology in school or trying to understand complex biological issues such as human health and ecological systems, the best approach is to incorporate both approaches.

But it's not always simply a matter of combining them. While there are times when drawing on both of them is the best way forward, there are other times when only one has a role to play, which one depending on the question one wants answered. Think about the illness of a family pet. To determine what precisely is the matter with the animal and how, hopefully, the illness can be treated, most people, including the two of us, would favor a conventional veterinary approach, one that is likely to draw principally on a mechanistic understanding of animals and how they function. But to attempt to understand in this way the role that a pet plays in a family is to miss something. The pet needs to be seen as an individual in its own right—the organismic approach—and in many households, it needs to be seen as part of the family, so that a systems approach is likely to be appropriate in understanding the effect that its illness has on the human members of the family.

Twentieth-century biology was more likely to err on the side of reductionism than holism, but there are now an increasing number of books that critique reductionist attempts to provide explanations about the operation of genes (Noble 2006; Alexander 2017; DeSalle & Tattersall 2018), hormones (Wiseman 2016), cells (Sharma 2015), the nervous system (Rose & Rose 2013, 2016), entire ecosystems (Gilbert & Gilbert 2019), and biology as a whole (Voit 2016). Exact predictions and all-encompassing explanations are relatively rare in biology, at least at the whole-organism level and above. Most biological processes are affected, directly or indirectly, by such a complex network of causal factors that exact predictions that turn out to be confirmed are simply not found (Watts & Reiss 2017). Biological predictions therefore tend to be probabilistic rather than exact for all but the most small-scale of predictions. Consider, for example, the behavior of potential prey (e.g., songbirds) when faced with a predator (e.g., a falcon). One might think that one can safely predict that in such a situation, songbirds scatter—except that while this is often the case, they sometimes freeze, and they occasionally attack back. At the whole organism level, predictions that have a very high degree of accuracy are all too often somewhat banal—along the lines of green plants grow less if kept in the dark.

This is not a counsel of despair! Biology can and does make predictions—it's just that things are generally more complicated than they are in physics and chemistry for the simple reason that biology has all the complications of those two sciences and a whole lot more beside. One just has to think how the science of COVID-19 has changed since it first appeared. What we have attempted to show is the way in which much of biology benefits from considering both a reductionist and a holistic approach—and this is true even for what might, at first, be thought to be issues where reductionism would suffice, such as the causes and effects of sickle cell anemia. Of course, there is nothing wrong with individual research biologists or teams of biologists choosing one approach over the other—we all have our own personalities and intellectual predispositions. It's rather that funders of biological research and those involved in formulating policy and large-scale programs of action should keep in mind that we need both approaches. That is particularly true of the "big" questions in biology—such as ones to do with climate change, agriculture, human health, and brain science.

Our argument applies not just to the science of biology but also to how biology is taught and communicated more generally—hence Chapter 9, "Biology Education"—and to how it interacts with other disciplines—see Chapter 8, "God and the New Biology." Finally, the complexity of biology means that there is considerable scope for each person, whether a scientist or not, to see the world in a way that not everyone else does. The search for a *unified* framework in biology is therefore not likely to be entirely successful. As we have argued, alternative ways of understanding biological systems are not alternative hypotheses where just one is verified and the others refuted. Rather, such ways operate as metaphors—each shines light on the issue at hand, but which one prefers may say as much about oneself as about the issue.

REFERENCES

ACKNOWLEDGMENTS

INDEX

REFERENCES

Adams, D. 2018. "Rudolf Steiner on traditional childhood illnesses and vaccines." *Our Spirit,* March 1, 2018. https://neoanthroposophy.com/2018/03/01/rudolf-steiner-on-traditional-childhood-illnesses-and-vaccines/.

Agricola, G. [1556] 1950. *De Re Metallica.* Translated by H. C. Hoover & L. C. Hoover. New York: Dover.

Akinyanju, O. O., A. I. Otaigbe, & M. O. Ibidapo. 2005. "Outcome of holistic care in Nigerian patients with sickle cell anaemia." *Clinical & Laboratory Haematology* 27, no. 3: 195–199.

Alexander, D. R. 2017. *Genes, Determinism and God.* Cambridge: Cambridge University Press.

Alfano, M. 2020. "Leadership of the journal Philosophical Psychology must answer for publishing race science." https://www.change.org/p/philosophers-and-other-researchers-in-humanities-and-social-sciences-editorial-board-of-the-journal-philosophical-psychology-must-resign.

Allen, E., B. Beckwith, J. Beckwith, S. Chorover, D. Culver, M. Duncan, S. J. Gould, R. Hubbard, H. Inouye, A. Leeds, R. Lewontin, C. Madansky, L. Miller, R. Pyeritz, M. Rosenthal, H. Schreier et al. 1975. "Letter to the editor." *New York Review of Books,* sec. 22, o18, pp. 43–44.

Allen, G. E. 1978. *Thomas Hunt Morgan: The Man and His Science.* Princeton, NJ: Princeton University Press.

Ameh, S. J., F. D. Tarfa, & B. U. Ebeshi. 2012. "Traditional herbal management of sickle cell anemia: Lessons from Nigeria." *Anemia* 2012: 607436.

American Psychiatric Association. 2013. *Diagnostic and Statistical Manual of Mental Disorders (DSM-5).* 5th ed. Washington, DC: American Psychiatric Publishing.

American Psychological Association. 2009. "Report of the APA Task Force on Gender Identity and Gender Variance." https://www.apa.org/pi/lgbt/resources/policy/gender-identity-report.pdf.

Anselm, St. 1903. *Anselm: Proslogium, Monologium, An Appendix on Behalf of the Fool by Gaunilon; and Cur Deus Homo.* Edited by S. N. Deane. Chicago: Open Court.

Appel, J. M. 2019. "Deaf parents want Deaf baby: Bioethicist weighs in." *MedPage Today,* October 11, 2019. https://www.medpagetoday.com/publichealthpolicy/ethics/82690.

Aquinas, St., T. 1952. *Summa Theologica, I.* London: Burns, Oates and Washbourne.

Aquinas, St., T. 1981. *Summa Theologica.* Translators Fathers of the English Dominican Province. London: Christian Classics.

Arthur, W. 2021. *Understanding Evo-Devo.* Cambridge: Cambridge University Press.

Atkins, L., & S. Michie. 2013. "Changing eating behaviour." *Nutrition Bulletin* 38: 30–35.

Augustine, St. 1982. *The Literal Meaning of Genesis.* Translated by J. H. Taylor. New York: Newman.

Australian Institute of Health and Welfare. 2020. "Deaths in Australia." https://www.aihw.gov.au/getmedia/e0df638c-0927-485a-b41c-fcedbd25072f/Deaths-in-Australia.pdf.aspx?inline=true.

Babkin, I. V., & I. N. Babkina. 2015. "The origin of the variola virus." *Viruses* 7, no. 3: 1100–1112.

Baedke, J. 2019. "O organism, where art thou? Old and new challenges for organism-centered biology." *Journal of the History of Biology* 52: 293–324.

Barbour, I. G. 1969. "Teilhard's process metaphysics." *The Journal of Religion* 49: 136–159.

Barbour, I. G. 1997. *Religion and Science: Historical and Contemporary Issues.* San Francisco: Harper.

Barnes, J., ed. 1984. *The Complete Works of Aristotle.* Princeton, NJ: Princeton University Press.

Barnshaw, J. 2008. "Race." In *Encyclopedia of Race, Ethnicity, and Society, Volume 1,* edited by R. T. Schaefer, 1091–1093. Thousand Oaks, CA: Sage.

Bateson, P., N. Cartwright, J. Dupré, K. Laland, & D. Noble. 2017. "New trends in evolutionary biology: Biological, philosophical and social science perspectives." *Interface Focus* 7, no. 5.

BBC. 2010. "Health Check: The boy who was raised a girl." *BBC,* November 23, 2010. https://www.bbc.co.uk/news/health-11814300.

BBC. 2020. "George Floyd: What happened in the final moments of his life." *BBC,* July 16, 2020. https://www.bbc.co.uk/news/world-us-canada-52861726.

Benjamin, R. 2019. *Race after Technology: Abolitionist Tools for the new Jim Code.* Cambridge: Polity.

Bergson, H. 1907. *L'évolution créatrice.* Paris: Alcan.

Bergson, H. 1911. *Creative Evolution.* New York: Holt.

Berry, T. 2009. *The Sacred Universe: Earth, Spirituality, and Religion in the Twenty-First Century.* Edited by E.-M. Tucker. New York: Columbia University Press.

Betts, H. C., M. N. Puttick, J. W. Clark, T. A. Williams, P. C. J. Donoghue, & D. Pisani. 2018. "Integrated genomic and fossil evidence illuminates life's early evolution and eukaryote origin." *Nature Ecology & Evolution* 2: 1556–1562.

Bivins, R. 2007. "'The English disease' or 'Asian rickets'? Medical responses to postcolonial immigration." *Bulletin of the History of Medicine* 81, no. 3: 533–568.
Blackless, M., A. Charuvastra, A. Derryck, A. Fausto-Sterling, K. Lauzanne, & E. Lee. 2000. "How sexually dimorphic are we? Review and synthesis." *American Journal of Human Biology* 12, no. 2: 151–166.
Blakemore, E. 2019. "Race and ethnicity: How are they different?" *National Geographic,* February 22, 2019. https://www.nationalgeographic.com/culture/article/race-ethnicity.
Boas, F. 1928. *Anthropology and Modern Life.* New York: Norton.
Boas, F. 1940. *Race, Language and Culture.* New York: Macmillan.
Boulton, J. G., P. M. Allen, & C. Bowman. 2015. *Embracing Complexity: Strategic Perspectives for an Age of Turbulence.* Oxford: Oxford University Press.
Bowler, P. J. 1989. *The Mendelian Revolution: The Emergence of Hereditarian Concepts in Modern Science and Society.* London: Athlone Press.
Boyle, R. [1688] 1966. "A disquisition about the final causes of natural things." In *The Works of Robert Boyle,* edited by T. Birch, 5th ed., 392–444. Hildesheim, Germany: Georg Olms.
Boyle, R. 1996. *A Free Enquiry into the Vulgarly Received Notion of Nature.* Edited by E. B. Davis & M. Hunter. Cambridge: Cambridge University Press.
Brown, W., N. Murphy, & H. N. Malony. 1998. *Whatever Happened to the Soul? Scientific and Theological Portraits of Human Nature.* Minneapolis, MN: Fortress Press.
Brugnara, C. 2018. "Sickle cell dehydration: Pathophysiology and therapeutic applications." *Clinical Hemorheology and Microcirculation* 68, no. 2–3: 187–204.
Buckland, W. 1823. *Reliquiae Diluvianae.* London: John Murray.
Burbidge, D., A. Briggs, & M. J. Reiss. 2020. *Citizenship in a Networked Age: An Agenda for Rebuilding Our Civic Ideals.* Oxford: University of Oxford. https://citizenshipinanetworkedage.org.
Butler, M. G., M. P. Walzak, W. G. Sanger, & C. T. Todd. 1983. "A possible etiology of the infertile 46XX male subject." *The Journal of Urology* 130, no. 1: 154–156.
Cain, A. J. 1954. *Animal Species and Their Evolution.* London: Hutchinson.
Callaghan, G. N. 2009. *Between XX and XY: Intersexuality and the Myth of Two Sexes.* Chicago: Chicago Review Press.
Cannon, W. B. 1931. *The Wisdom of the Body.* Cambridge, MA: Harvard University Press.
Canon, G. 2021. "California couple whose gender-reveal party sparked a wildfire charged with 30 crimes." *The Guardian,* July 21, 2021. https://www.theguardian.com/us-news/2021/jul/21/couple-gender-reveal-party-wildfire-charged.
Capra, F. 1987. "Deep ecology: A new paradigm." In *Deep Ecology for the 21st Century,* edited by G. Sessions, 19–25. Boston: Shambhala.
Carothers, B. J., & H. T. Reis. 2012. "Men and women are from Earth: Examining the latent structure of gender." *Journal of Personality and Social Psychology* 104, no. 2: 385–407.
Carroll, A., D. Stokes, & A. Darley. 2021. "Use of complexity theory in health and social care: A scoping review protocol." *BMJ Open* 11: e047633.

Carroll, S. B. 2005. *Endless Forms Most Beautiful: The New Science of Evo Devo.* New York: Norton.
Carroll, S. B., J. K. Grenier, & S. D. Weatherbee. 2021. *From DNA to Diversity: Molecular Genetics and the Evolution of Animal Design.* Oxford: Blackwell.
Carson, R. [1950] 2018. *The Sea Around Us.* Oxford: Oxford University Press.
Carson, R. 1955. *The Edge of the Sea.* Boston: Houghton Mifflin.
Carson, R. 1962. *Silent Spring.* New York: Houghton Mifflin.
Centers for Disease Control and Prevention. 2021. "Healthy weight, nutrition, and physical activity: The health effects of overweight and obesity." https://www.cdc.gov/healthyweight/effects/index.html.
Charache, S., M. L. Terrin, R. D. Moore, G. J. Dover, F. B. Barton, S. V. Eckert, R. P. McMahon, & D. R. Bonds. 1995. "Effect of hydroxyurea on the frequency of painful crises in sickle cell anemia." *New England Journal of Medicine* 332, no. 20: 1317–1322.
Chien, W. T., S. F. Leung, F. K. Yeung, & W. K. Wong. 2013. "Current approaches to treatments for schizophrenia spectrum disorders, part II: Psychosocial interventions and patient-focused perspectives in psychiatric care." *Neuropsychiatric Disease and Treatment* 9: 1463–1481.
Clark, R. W. 1960. *Sir Julian Huxley, F.R.S.* London: Phoenix House.
Clayton, P. 2006. *Mind and Emergence: From Quantum to Consciousness.* Oxford: Oxford University Press.
Clifford, W. K. [1874] 1901. "Body and mind (from *Fortnightly Review*)." In *Lectures and Essays of the Late William Kingdom Clifford,* edited by L. Stephen & F. Pollock, Vol. 2, 1–51. London: Macmillan.
CNN. 1999. "Grisham ranks as topselling author of decade." https://web.archive.org/web/20120908181659/http://articles.cnn.com/1999-12-31/entertainment/1990.sellers_1_book-sales-cumulative-sales-copies?_s=PM:books.
Cobb, J. 1991. *Matters of Life and Death.* Louisville, KY: Westminster.
Cofnas, N. 2020. "Research on group differences in intelligence: A defense of free inquiry." *Philosophical Psychology* 33, no. 1: 125–147.
Colapinto, J. 2000. *As Nature Made Him: The Boy Who Was Raised as a Girl.* New York: Harper Perennial.
Coleman, W. 1964. *Georges Cuvier Zoologist: A Study in the History of Evolution Theory.* Cambridge, MA: Harvard University Press.
Conway Morris, S. 2003. *Life's Solution: Inevitable Humans in a Lonely Universe.* Cambridge: Cambridge University Press.
Cooper, A. 2019. "Hear me out: Hearing each other for the first time: The implications of cochlear implant activation." *Missouri Medicine* 116, no. 6: 469–471.
Cooper, J. M., ed. 1997. *Plato: Complete Works.* Indianapolis: Hackett.
Coyne, J. A. 2012. "Another philosopher proclaims a nonexistent 'crisis' in evolutionary biology." https://whyevolutionistrue.com/2012-09/07/another-philosopher-proclaims-a-nonexistent-crisis-in-evolutionary-biology/.
Coyne, J. A. 2015. *Faith Versus Fact: Why Science and Religion Are Incompatible.* New York: Viking.
Coyne, J. A. 2017. "More dumb claims that environmental epigenetics will completely revise our view of evolution." https://whyevolutionistrue.com/2017/10/27/more

-dumb-theorizing-that-epigenetics-as-lamarckian-inheritance-will-completely -revise-our-view-of-evolution/.

Coyne, J. A. 2019. "Epigenetics: The return of Lamarck? Not so fast!" https://whyevolutionistrue.com/2018/08/26/epigenetics-the-return-of-lamarck-not-so -fast/.

Cuvier, G. [1813] 1822. *Theory of the Earth*. Edited by Robert Jameson, 4th ed. Edinburgh: William Blackwood.

Daaboul, J., & J. E. Frader. 2001. "Ethics and the management of the patient with intersex: A middle way." *Journal of Pediatric Endocrinology and Metabolism* 14, no. 9: 1575–1583.

Dahlgren, G., & M. Whitehead. 1991. *Policies and Strategies to Promote Social Equity in Health*. Stockholm: Institute for Futures Studies.

Darwin, C. 1859. *On the Origin of Species by Means of Natural Selection, or the Preservation of Favoured Races in the Struggle for Life*. London: John Murray.

Darwin, C. 1861. *Origin of Species, Third Edition*. London: John Murray.

Darwin, C. 1868. *The Variation of Animals and Plants Under Domestication*. London: John Murray.

Darwin, C. 1871. *The Descent of Man, and Selection in Relation to Sex*. London: John Murray.

Darwin, C. 1881. *The Formation of Vegetable Mould, Through the Action of Worms, with Observations on their Habits*. London: John Murray.

Darwin, C. 1985–. *The Correspondence of Charles Darwin*. 30 Vols. Cambridge: Cambridge University Press.

Darwin, C. 1987. *Charles Darwin's Notebooks, 1836–1844*. Edited by P. H. Barrett, P. J. Gautrey, S. Herbert, D. Kohn, & S. Smith. Ithaca, NY: Cornell University Press.

Darwin, E. [1794–1796] 1801. *Zoonomia; or, The Laws of Organic Life*. 3rd ed. London: J. Johnson.

Davy, Z. 2021. *Sex/Gender and Self-determination: Policy Developments in Law, Health and Pedagogical Contexts*. Bristol: Bristol University Press.

Dawkins, R. 1976. *The Selfish Gene*. Oxford: Oxford University Press.

Dawkins, R. 1982. *The Extended Phenotype: The Gene as the Unit of Selection*. Oxford: W. H. Freeman.

Dawkins, R. 1983. "Universal Darwinism." In *Evolution from Molecules to Men*, edited by D. S. Bendall, 403–425. Cambridge: Cambridge University Press.

Dawkins, R. 1986. *The Blind Watchmaker*. New York: Norton.

Dawkins, R. 2006. *The God Delusion*. New York: Houghton, Mifflin, Harcourt.

Dawkins, R., & J. R. Krebs. 1979. "Arms races between and within species." *Proceedings of the Royal Society of London B* 205: 489–511.

de Beer, G. R. 1940. *Embryos and Ancestors*. Oxford: Oxford University Press.

Dea, S. 2016. *Beyond the Binary: Thinking about Sex and Gender*. Peterborough, Canada: Broadview Press.

Dennett, D. C. 1984. *Elbow Room: The Varieties of Free Will Worth Wanting*. Cambridge, MA: MIT Press.

Dennett, D. C., & R. Winston. 2008. "Is religion a threat to rationality and science?" *The Guardian,* April 22, 2008.

Department of the Prime Minister and Cabinet. 2017. "Closing the gap: Prime Minister's report 2017." https://www.niaa.gov.au/sites/default/files/publications/ctg-report-2017.pdf.

DeSalle, R., & I. Tattersall. 2018. *Troublesome Science: The Misuse of Genetics and Genomics in Understanding Race.* New York: Columbia University Press.

Descartes, R. [1637] 1964. "Discourse on Method." In *Philosophical Essays,* 1–57. Indianapolis: Bobbs-Merrill.

Descartes, R. [1642] 1964. "Meditations." In *Philosophical Essays,* 59–143. Indianapolis: Bobbs-Merrill.

Descartes, R. [1701] 1964. "Rules for the direction of the mind." In *Philosophical Essays,* 145–236. Indianapolis: Bobbs-Merrill.

Dickens, B. M. 2018. "Management of intersex newborns: Legal and ethical developments." *International Journal of Gynaecology and Obstetrics* 143, no. 2: 255–259.

Dijksterhuis, E. J. 1961. *The Mechanization of the World Picture.* Oxford: Oxford University Press.

Dixit, R., S. Nettem, S. S. Madan, H. H. K. Soe, A. B. L. Abas, L. D. Vance, & P. J. Stover. 2018. "Folate supplementation in people with sickle cell disease." *Cochrane Database of Systematic Reviews* 3: CD011130.

Dixson, A. F., & B. J. Dixson. 2011. "Venus figurines of the European Paleolithic: Symbols of fertility or attractiveness?" *Journal of Anthropology* 2011: 569120.

Dobzhansky, T. 1937. *Genetics and the Origin of Species.* New York: Columbia University Press.

Dobzhansky, T. 1943. "Temporal changes in the composition of populations of *Drosophila pseudoobscura* in different environments." *Genetics* 28: 162–186.

Donovan, B. M., M. Weindling, & D. M. Lee. 2020. "From basic to humane genomics literacy: How different types of genetics curricula could influence anti-essentialist understandings of race." *Science & Education* 29: 1479–1511.

Doolittle, W. F. 1981. "Is nature really motherly?" *CoEvolution* 29: 58–62.

Downing, S. 2016. "'Boys just don't do that': Meet the Castro's pioneering male nurse midwife." *hoodline,* September 26, 2016. https://hoodline.com/2016/09/boys-just-don-t-do-that-meet-the-castro-s-pioneering-male-nurse-midwife/.

Draper, J. W. 1875. *History of the Conflict Between Religion and Science.* New York: Appleton.

Duncan, D., ed. 1908. *Life and Letters of Herbert Spencer.* London: Williams and Norgate.

Duncan, R. 2018. *Nature in Mind: Systematic Thinking and Imagination in Ecopsychology and Mental Health.* London: Routledge.

Dupré, J. 2003. *Darwin's Legacy: What Evolution Means Today.* Oxford: Oxford University Press.

Dupré, J. 2010. "The conditions for existence." *American Scientist* 98: 170.

Dupré, J. 2012a. *Processes of Life: Essays in the Philosophy of Biology.* Oxford: Oxford University Press.

Dupré, J. 2012b. "Evolutionary theory's welcome crisis." https://www.project-syndicate.org/commentary/evolutionary-theory-s-welcome-crisis-by-john-dupre?barrier=accesspaylog.

Dupré, J. 2017. "The metaphysics of evolution." *Interface Focus* 7, no. 5. https://doi.org/10.1098/rsfs.2016.0148.

Durbach, N. 2000. "They might as well brand us: Working class resistance to compulsory vaccination in Victorian England." *The Society for the Social History of Medicine* 13, no. 1: 45–62.

Ecklund, E. H., & D. R. Johnson. 2021. *Varieties of Atheism in Science*. New York: Oxford University Press.

Edwards, K. M., J. M. Hackell, & The Committee on Infectious Diseases, The Committee on Practice and Ambulatory Medicine. 2016. "Countering vaccine hesitancy." *Pediatrics* 138, no. 3: e20162146.

Ehrhardt, A. A. 2007. "John Money, Ph.D." *The Journal of Sex Research* 44, no. 3: 223–224.

Eldredge, N., & S. J. Gould. 1972. "Punctuated equilibria: An alternative to phyletic gradualism." In *Models in Paleobiology*, edited by T. J. M. Schopf, 82–115. San Francisco: Freeman, Cooper.

Eliot, L. 2009. *Pink Brain, Blue Brain: How Small Differences Grow into Troublesome Gaps*. Oxford: Oneworld.

Eliot, L. 2019. "Neurosexism: The myth that men and women have different brains." *Nature* 566: 453–454.

Epperly, B. G. 2011. *Process Theology: A Guide for the Perplexed*. London: T&T Clark.

Ernst, K., & E. T. Jacobs. 2012. "Implications of philosophical and personal belief exemptions on re-emergence of vaccine-preventable disease: The role of spatial clustering in under-vaccination." *Human Vaccines & Immunotherapeutics* 8, no. 6, 838–841.

Esbjorn-Hargens, S., & M. E. Zimmerman. 2009. *Integral Ecology: Uniting Multiple Perspectives on the Natural World*. Boston: Integral Books.

Evangelical Lutheran Church in America. 1996. "Basis for our caring." In *This Sacred Earth: Religion, Nature, Environment*, edited by R. S. Gottlieb, 235–252. New York: Routledge.

Evans, M. A., & H. E. Evans. 1970. *William Morton Wheeler, Biologist*. Cambridge, MA: Harvard University Press.

Faber, J. W. 2020. "We built this: Consequences of New Deal Era intervention in America's racial geography." *American Sociological Review* 85, no. 5: 739–775.

Farm Animal Welfare Council. 2009. *Farm Animal Welfare in Great Britain: Past, Present and Future*. London: Farm Animal Welfare Council. https://assets.publishing.service.gov.uk/government/uploads/system/uploads/attachment_data/file/319292/Farm_Animal_Welfare_in_Great_Britain_-_Past__Present_and_Future.pdf.

Fauquet, C. M., R. W. Briddon, J. K. Brown, E. Moriones, J. Stanley, M. Zerbini, & X. Zhou. 2008. "Geminivirus strain demarcation and nomenclature." *Archives of Virology* 153: 783–821.

Fausto-Sterling, A. 2000. *Sexing the Body*. New York: Basic Books.
Fisher, R. A. 1930. *The Genetical Theory of Natural Selection*. Oxford: Oxford University Press.
Fisher, R. A., & E. B. Ford. [1947] 1974. "The spread of a gene in natural conditions in a colony of the moth *Panaxia dominula*." *Heredity* 1: 143–74.
Fodor, J. 2007. "Why pigs don't have wings. The case against natural selection." *London Review of Books* 29, no. 20, October 18, 2007.
Ford, E. B. 1964. *Ecological Genetics*. London: Methuen.
Foresight. 2007. "Tackling obesities: Future Choices—Project report." https://assets.publishing.service.gov.uk/government/uploads/system/uploads/attachment_data/file/287937/07-1184x-tackling-obesities-future-choices-report.pdf.
Foulkes, M. 2021. "How the spiritual 'Waldorf' movement is connected to German vaccine scepticism." *The Local Germany*, November 23, 2021. https://www.thelocal.de/20211123/how-a-spiritual-movement-is-connected-to-german-vaccine-scepticism/.
Francis (Pope). 2015. *Encyclical on Climate Change and Inequality: On Care for Our Common Home*. New York: Melville House.
Fry, D. P., ed. 2013. *War, Peace, and Human Nature: The Convergence of Evolutionary and Cultural Views*. Oxford: Oxford University Press.
Futuyma, D. J. 2017. "Evolutionary biology today and the call for an extended synthesis." *Interface Focus* 7, no. 5. https://doi.org/10.1098/rsfs.2016.0145.
Gaetano, P. 2021. "David Reimer and John Money gender reassignment controversy: The John / Joan case." http://embryo.asu.edu/handle/10776/13009.
Gallaudet University. 2021. "Fast facts." https://www.gallaudet.edu/about/news-and-media/fast-facts.
Gammon, J. 2015. "Multi-country study finds body image improves with age." https://today.yougov.com/topics/lifestyle/articles-reports/2015/07/21/multi-country-study-find-body-image-improves-age.
Ganguly, P., A. Soliman, & A. A. Moustafa. 2018. "Holistic management of schizophrenia symptoms using pharmacological and non-pharmacological treatment." *Frontiers in Public Health* 6: 166.
Gare, A. 2002. "The roots of postmodernism: Schelling, process philosophy and poststructuralism." In *Process and Difference: Between Cosmological and Poststructuralist Postmodernisms*, edited by C. Keller & A. Daniell, 31–54. Albany, NY: SUNY Press.
Gayon, J. 2013. "Darwin and Darwinism in France after 1900." In *The Cambridge Encyclopedia of Darwin and Evolutionary Thought*, edited by M. Ruse, 300–312. Cambridge: Cambridge University Press.
Gianno, R. 2004. "'Women are not brave enough'; Semelai male midwives in the context of Southeast Asian cultures." *Bijdragen tot de Taal-, Land- en Volkenkunde* 160, no. 1: 31–71.
Gibson, A. 2013. "Edward O. Wilson and the organicist tradition." *Journal of the History of Biology* 46: 599–630.
Gilbert, F., & H. Gilbert. 2019. *Conservation: A People-Centred Approach*. Oxford: Oxford University Press.

Gilbert, S. 2006. "The generation of novelty: The province of developmental biology." *Biological Theory* 1: 209–212.

Gilissen, M. G. R., M.-C. P. J. Knippels, & W. R. van Joolingen. 2020. "Bringing systems thinking into the classroom." *International Journal of Science Education* 42, no. 8: 1253–1280.

Gilkey, L. B. 1985. *Creationism on Trial: Evolution and God at Little Rock*. Minneapolis, MN: Winston Press.

Glasgow, J., S. Haslanger, C. Jeffers, & Q. Spencer. 2019. *What is Race? Four Philosophical Views*. New York: Oxford University Press.

Godziewski, C. 2020. "Obesity strategy: Policies placing responsibility on individuals don't work—so why does the government keep using them?" *The Conversation,* August 18, 2020. https://theconversation.com/obesity-strategy-policies-placing-responsibility-on-individuals-dont-work-so-why-does-the-government-keep-using-them-144310?utm_medium=email&utm_campaign=Latest%20from%20The%20Conversation%20for%20August%2018%202020&utm_content=Latest%20from%20The%20Conversation%20for%20August%2018%202020+CID_e6fb346c8f854348a02f73bdca916834&utm_source=campaign_monitor_uk&utm_term=Obesity%20strategy%20policies%20placing%20responsibility%20on%20individuals%20dont%20work%20%20so%20why%20does%20the%20government%20keep%20using%20them.

Goethe, J. W. [1790] 1946. "On the metamorphosis of plants." In *Goethe's Botany, Chronica Botanica,* edited by A. Arber, 63–126. London: William Dawson and Sons.

Golding, W. 1983. "The Nobel Prize in literature." https://www.nobelprize.org/prizes/literature/1983/golding/facts/.

Goodman, A. H., Y. T. Moses, & J. L. Jones. 2019. *Race: Are We So Different?* 2nd ed. Hoboken, NJ: Wiley Blackwell.

Goodman, J., & F. Carmichael. 2020. "Coronavirus: Bill Gates 'microchip' conspiracy theory and other vaccine claims fact-checked." *BBC,* May 29, 2020. https://www.bbc.co.uk/news/52847648.

Goodwin, B. 2001. *How the Leopard Changed Its Spots*. 2nd ed. Princeton, NJ: Princeton University Press.

Gould, S. J. 1977. *Ontogeny and Phylogeny*. Cambridge, MA: Belknap Press.

Gould, S. J. 1981. *The Mismeasure of Man*. London: Norton.

Gould, S. J. 1982. "Darwinism and the expansion of evolutionary theory." *Science* 216: 380–387.

Gould, S. J. 1988. "On replacing the idea of progress with an operational notion of directionality." In *Evolutionary Progress,* edited by M. H. Nitecki, 319–338. Chicago: University of Chicago Press.

Gould, S. J. 1999. *Rocks of Ages: Science and Religion in the Fullness of Life*. New York: Ballantine.

Gould, S. J., & R. C. Lewontin. 1979. "The spandrels of San Marco and the Panglossian paradigm: A critique of the adaptationist programme." *Proceedings of the Royal Society of London, Series B: Biological Sciences* 205: 581–598.

Grant, P. R., & B. R. Grant. 2002. "Unpredictable evolution in a 30-year study of Darwin's finches." *Science* 296: 707–711.

Grant, P. R., & B. R. Grant. 2014. *40 Years of Evolution: Darwin's Finches on Daphne Major Island.* Princeton, NJ: Princeton University Press.

Gray, J. 1992. *Men Are from Mars, Women Are from Venus: A Practical Guide for Improving Communication and Getting What You Want in Your Relationships.* New York: Harper.

Greenhalgh, T., & C. Papoutsi. 2018. "Studying complexity in health services research: Desperately seeking an overdue paradigm shift." *BMC Medicine* 16: 95.

Greenwood, B. 2014. "The contribution of vaccination to global health: Past, present and future." *Philosophical Transactions of the Royal Society of London. Series B, Biological Sciences* 369, no. 1645: 20130433.

Gregersen, N. H. 2017. "The exploration of ecospace: Extending or supplementing the neo-Darwinian paradigm?" *Zygon* 52: 561–586.

Gwynn, H. 2020. "Permaculture chooses farming practices with an eye to a sustainable future." *Tallahassee Democrat,* https://www.tallahassee.com/story/life/causes/2020-08/24/permaculture-chooses-farming-practices-eye-sustainablity/3407851001/.

Haeckel, E. 1896. *The Evolution of Man.* 2 Vols. New York: Appleton.

Halberstam, J. 2018. *Trans*: A Quick and Quirky Account of Gender Variability.* Oakland: University of California Press.

Haldane, J. B. S. 1932. *The Causes of Evolution.* New York: Cornell University Press.

Haldane, J. S. 1935. *The Philosophy of a Biologist.* Oxford: Oxford University Press.

Hall, A. R. 1954. *The Scientific Revolution 1500–1800: The Formation of the Modern Scientific Attitude.* London: Longman, Green.

Hallam, A. 1973. *A Revolution in the Earth Sciences.* Oxford: Oxford University Press.

Hamilton, W. D. 1964a. "The genetical evolution of social behaviour I." *Journal of Theoretical Biology* 7: 1–16.

Hamilton, W. D. 1964b. "The genetical evolution of social behaviour II." *Journal of Theoretical Biology* 7: 17–32.

Hamilton, W. D., & T. Lenton. 2005. "Spora and Gaia: How microbes fly with their clouds." In *Narrow Roads of Gene Land: Volume 3. Last Words,* 271–289. Oxford: Oxford University Press.

Hardin, J., R. Numbers, & R. A. Binzley, eds. 2018. *The Warfare Between Science and Religion (The Idea that Wouldn't Die).* Baltimore: Johns Hopkins University Press.

Hardy, T. 1994. *Collected Poems.* Ware, UK: Wordsworth Poetry Library.

Harmsen, N. 2019. "The inventor of the gender reveal party now has a daughter who prefers suits." *Today's Parent,* July 31, 2019. https://www.todaysparent.com/blogs/trending/gender-reveal-party-inventors-daughter-prefers-suits/.

Harrington, A. 1996. *Reenchanted Science: Holism in German Culture from Wilhelm II to Hitler.* Princeton, NJ: Princeton University Press.

Harrison, H. S. 1936. "Concerning human progress." *Journal of the Royal Anthropological Institute* 66: 1.

Harvey, A. 2016. "The list that can take your life." https://www.huffpost.com/entry/the-list-that-can-take-your-life_b_57eae82ce4b07f20daa0fd51?guce_referrer=aH

R0cHM6Ly93d3cuZ29vZ2xlLmNvbS8&guce_referrer_sig=AQAAAKGH5eIW3DqMjFy9T4mfkKHmJtMHCbpnHIlU7Y5VN4kWCfU5_oFmFHiqr-L-zyLNY-UQxH5GmYxqnUbq0Jvu36oOAxZzImO7vIoNZW_pRCxldChZ_ttKob_f5F3nHPS62T-D2Vh7FBj6qcNm7DvSe24nlBAyNHy9nX-xxSUOWbpM&guccounter=2.

Harvey, P. 1990. *An Introduction to Buddhism: Teachings, History and Practices.* Cambridge: Cambridge University Press.

Hattie, J., V. Bustamante, J. T. Almarode, D. Fisher, & N. Frey. 2020. *Great Teaching by Design: From Intention to Implementation in the Visible Learning Classroom.* Thousand Oaks, CA: Corwin.

Healthline.com. 2021. "Obesity." https://www.healthline.com/health/obesity.

Heidegger, M. 1959. *An Introduction to Metaphysics.* New Haven, CT: Yale University Press.

Helm, A. B. 2016. "'3 Black Teens' Google search sparks outrage." *The Root,* June 12, 2016. https://www.theroot.com/3-black-teens-google-search-sparks-outrage-1790855635.

Henderson, L. J. 1917. *The Order of Nature.* Cambridge, MA: Harvard University Press.

Herrnstein, R. J., & C. Murray. 1994. *The Bell Curve: Intelligence and Class Structure in American Life.* New York: Free Press.

Herschberger, R. 1948. *Adam's Rib.* New York: Pelligrini & Cudaby.

Herschel, J. F. W. 1830. *Preliminary Discourse on the Study of Natural Philosophy.* London: Longman, Rees, Orme, Brown, Green, & Longman.

Hertler, S. C., A. J. Figueredo, & M. Peñaherrera-Aguirre. 2020. *Multilevel Selection: Theoretical Foundations, Historical Examples, and Empirical Evidence.* London: Palgrave Macmillan.

Hirt, V., I. Schalinski, & B. Rockstroh. 2019. "Decoding the impact of adverse childhood experiences on the progression of schizophrenia." *Mental Health & Prevention* 13: 82–91.

Hoban, M. D., D. Lumaquin, C. Y. Kuo, Z. Romero, J. Long, M. Ho, C. S. Young, M. Mojadidi, S. Fitz-Gibbon, A. R. Cooper, G. R. Lill, F. Urbinati, B. Campo-Fernandez, C. F. Bjurstrom, M. Pellegrini, R. P. Hollis, & D. B. Kohn. 2016. "CRISPR/Cas9-mediated correction of the sickle mutation in human CD34+ cells." *Molecular Therapy: The Journal of the American Society of Gene Therapy* 24: 1561–1569.

Hölldobler, B., & E. O. Wilson. 2008. *The Superorganism: The Beauty, Elegance, and Strangeness of Insect Societies.* New York: Norton.

Hopson, J. A. 1977. "Relative brain size and behavior in archosaurian reptiles." *Annual Review of Ecology and Systematics* 8: 429–448.

Hublin, J., A. Ben-Ncer, S. Bailey, S. E. Freidline, S. Neubauer, M. M. Skinner, I. Bergmann, A. Le Cabec, S. Benazzi, K. Harvati, & P. Gunz. 2017. "New fossils from Jebel Irhoud, Morocco and the pan-African origin of *Homo sapiens.*" *Nature* 546, no. 7657: 289–292.

Hull, D. L. 1974. *The Philosophy of Biological Science.* Englewood Cliffs: Prentice-Hall.

Hull, D. L. 1976. "Are species really individuals?" *Systematic Zoology* 25: 174–191.

Hull, D. L. 1978. "A matter of individuality." *Philosophy of Science* 45: 335–360.

Hume, D. [1779] 1990. *Dialogues Concerning Natural Religion*. Edited by M. Bell. London: Penguin.
Hursthouse, R. 1999. *On Virtue Ethics*. Oxford: Oxford University Press.
Hutchinson, G. E. 1948. "Circular causal systems in ecology." *Annals of the New York Academy of Science* 50: 221–246.
Hutton, J. 1788. "Theory of the Earth; or an investigation of the laws observable in the composition, dissolution, and restoration of land upon the globe." *Transactions of the Royal Society of Edinburgh* 1, no. 2: 209–304.
Huxley, J. S. 1912. *The Individual in the Animal Kingdom*. Cambridge: Cambridge University Press.
Huxley, J. S. 1934. *If I Were Dictator*. New York: Harper.
Huxley, J. S. 1942. *Evolution: The Modern Synthesis*. London: Allen and Unwin.
Huxley, J. S. 1943. *TVA: Adventure in Planning*. London: Scientific Book Club.
Huxley, J. S. 1948. *UNESCO: Its Purpose and Its Philosophy*. Washington, D.C.: Public Affairs Press.
Insel, T. 2010. "Rethinking schizophrenia." *Nature* 468: 187–193.
Intersex Human Rights Australia. 2013. "Intersex population figures." https://ihra.org.au/16601/intersex-numbers/.
Jablonka, E. 2017. "The evolutionary implications of epigenetic inheritance." *Interface Focus* 7, no. 5. https://www.ncbi.nlm.nih.gov/pmc/articles/PMC5566804/.
Jablonka, E., & M. Lamb. 2020. *Inheritance Systems and the Extended Synthesis (Elements in the Philosophy of Biology)*. Cambridge: Cambridge University Press.
Jackson, J. P., Jr., & D. J. Depew. 2017. *Darwinism, Democracy, and Race: American Anthropology and Evolutionary Biology in the Twentieth Century*. London: Routledge.
Jain, H. K. 2010. *Green Revolution: History, Impact and Future*. Houston: Studium Press.
Jamme, H.-T. W., D. Bahl, & T. Banerjee. 2018. "Between 'broken windows' and the 'eyes on the street': Walking to school in inner city San Diego." *Journal of Environmental Psychology* 55: 121–138.
Jensen, A. R. 1969. "How much can we boost IQ and scholastic achievement?" *Harvard Educational Review* 39, no. 1: 1–123.
Jerison, H. 1973. *Evolution of the Brain and Intelligence*. New York: Academic Press.
Joel, D., Z. Berman, I. Tavor, N. Wexler, O. Gaber, Y. Stein, N. Shefi, J. Pool, S. Urchs, D. S. Margulies, F. Liem, J. Hänggi, L. Jäncke, & Y. Assaf. 2015. "Sex beyond the genitalia: The human brain mosaic." *Proceedings of the National Academy of Sciences* 112, no. 50: 15468–15473.
Johansson, M., T. Hartig, & H. Staats. 2011. "Psychological benefits of walking: Moderation by company and outdoor environment." *Applied Psychology: Health and Well-Being* 3: 261–280.
Johns Hopkins Medicine. 2021. "Chinese medicine." https://www.hopkinsmedicine.org/health/wellness-and-prevention/chinese-medicine.
Johnson, G. R. 1986. "Kin selection, socialization, and patriotism: an integrating theory." *Politics and the Life Sciences* 4: 127–140.

Kant, I. [1790] 1928. *The Critique of Teleological Judgement*. Translated by J. C. Meredith. Oxford: Oxford University Press.

Kauffman, S. A. 1993. *The Origins of Order: Self-Organization and Selection in Evolution*. Oxford: Oxford University Press.

Kauffman, S. A. 1995. *At Home in the Universe: The Search for the Laws of Self-Organization and Complexity*. New York: Oxford University Press.

Kauffman, S. A. 2008. *Reinventing the Sacred: A New View of Science, Reason, and Religion*. New York: Basic Books.

Kepler, J. [1619] 1977. *The Harmony of the World*. Translated E. J. Aiton, A. M. Duncan, & J. V. Field. Philadelphia: American Philosophical Society.

Kessler, S. 1998. *Lessons from the Intersexed*. New Brunswick, NJ: Rutgers University Press.

Kettlewell, H. B. D. 1955. "Selection experiments on industrial melanism in the *Lepidoptera*." *Heredity* 9: 323–342.

King, D. 1996. "An interview with Professor Brian Goodwin." *GenEthics News* 11: 6–8.

Kinjo, K., T. Yoshida, Y. Kobori, H. Okada, E. Suzuki, T. Ogata, M. Miyado, & M. Fukami. 2020. "Random X chromosome inactivation in patients with Klinefelter syndrome." *Molecular and Cellular Pediatrics* 7, no. 1: 1. https://doi.org/10.1186/s40348-020-0093-x.

Kirby, W., & W. Spence. 1815–1828. *An Introduction to Entomology: or Elements of the Natural History of Insects*. London: Longman, Hurst, Reece, Orme, & Brown.

Klinefelter, Jr., H. F., E. C. Reifenstein, Jr., & F. Albright, Jr. 1942. "Syndrome characterized by gynecomastia, aspermatogenesis without A-Leydigism, and increased excretion of follicle-stimulating hormone." *The Journal of Clinical Endocrinology & Metabolism* 2, no. 11: 615–627.

Koch, A., C. Brierley, M. Maslin, & S. Lewis. 2019. "European colonization of the Americas killed 10 percent of world population and caused global cooling." *The Conversation*, January 31, 2019. https://www.pri.org/stories/2019-01-31/european-colonization-americas-killed-10-percent-world-population-and-caused.

Koopman, P., J. Gubbay, N. Vivian, P. Goodfellow, & R. Lovell-Badge. 1991. "Male development of chromosomally female mice transgenic for Sry." *Nature* 351, no. 6322: 117–121.

Koprowski, J., K. E. Munroe, & A. J. Edelman. 2016. "Gray not grey: The ecology of *Sciurus carolinensis* in their native range in North America." In *The Grey Squirrel: Ecology & Management of an Invasive Species in Europe*, edited by C. Shuttleworth, P. Lurz, & J. Gurnell, 1–18. Stoneleigh Park, UK: European Squirrel Initiative.

Kuhn, T. 1962. *The Structure of Scientific Revolutions*. Chicago: University of Chicago Press.

Kuhn, T. 1993. "Metaphor in science." In *Metaphor and Thought*, edited by A. Ortony, 2nd ed., 533–542. Cambridge: Cambridge University Press.

Laland, K., J. Odling-Smee, & J. Endler. 2017. "Niche construction, sources of selection and trait coevolution." *Interface Focus* 7, no. 5. https://royalsocietypublishing.org/doi/10.1098/rsfs.2016.0147.

Laland, K., T. Uller, M. Feldman, K. Sterelny, G. B. Müller, A. Moczek, E. Jablonka, J. Odling-Smee, G. A. Wray, H. E. Hoekstra, D. J. Futuyma, R. E. Lenski, T. F. C. Mackay, D. Schluter, & J. E. Strassmann. 2014. "Does evolutionary theory need a rethink?" *Nature* 514: 161–164.

Laland, K. N., T. Uller, M. W. Feldman, K. Sterelny, B. Müller, A. Moczek, E. Jablonka, & J. Odling-Smee 2015. "The extended evolutionary synthesis: Its structure, assumptions and predictions." *Proceedings of the Royal Society B* 282, no. 1813: 20151019. https://kevintshoemaker.github.io/EECB-703/Laland%20et%20al.%20-%202015%20-%20The%20extended%20evolutionary%20synthesis%20its%20structure.pdf.

Lamarck, J. B. 1815–1822. *Histoire naturelle des animaux sans vertèbres.* Paris: Verdiere.

Larson, E. J., & M. Ruse. 2017. *On Faith and Science.* New Haven, CT: Yale University Press.

Lawton, J. 2001. "Editorial: Earth system science." *Science* 292: 1965.

Leibniz, G. F. W. 1714. *Monadology and other Philosophical Essays.* New York: Bobbs-Merrill.

Leidenhag, J. 2019. "The revival of panpsychism and its relevance for the science-religion dialogue." *Theology and Science* 17: 90–106.

Lennox, J. G. 2001. *Aristotle's Philosophy of Biology.* Cambridge: Cambridge University Press.

Leopold, A. 1949. *A Sand County Almanac, and Sketches Here and There.* New York: Oxford University Press.

Leopold, A. 1979. "Some fundamentals of conservation in the Southwest." *Environmental Ethics* 1: 131–142.

Le Page, M. 2017. "Blind cave fish lost eyes by unexpected evolutionary process." *New Scientist,* October 12, 2017. https://www.newscientist.com/article/2150233-blind-cave-fish-lost-eyes-by-unexpected-evolutionary-process/.

Lewis, S. 2019. "The racial bias built into photography." *The New York Times,* April 25, 2019. https://www.nytimes.com/2019/04/25/lens/sarah-lewis-racial-bias-photography.html.

Lewontin, R. C. 1972. "The apportionment of human diversity." In *Evolutionary Biology,* edited by T. Dobzhansky, M. K. Hecht, & W. C. Steere, 381–398. New York: Springer.

Lewontin, R. C. 1974. *The Genetic Basis of Evolutionary Change.* New York: Columbia University Press. XXV in the Columbia Biology Series.

Lewontin, R. C. 1976. "Sociobiology—A caricature of Darwinism." *PSA: Proceedings of the Biennial Meeting of the Philosophy of Science Association* 2: 22–31.

Lewontin, R. C. 1983. "Gene, organism and environment." In *Evolution from Molecules to Men,* edited by D. S. Bendall, 275–285. Cambridge: Cambridge University Press.

Lewontin, R. C. 1991. *Biology as Ideology: The Doctrine of DNA.* Toronto: Anansi.

Lewontin, R. C., S. Rose, & L. J. Kamin. 1984. *Not in Our Genes: Biology, Ideology and Human Nature.* New York: Pantheon.

Leyser, O., & H. Wiseman. 2020. "Integrative biology: Parts, wholes, levels and systems." In *Rethinking Biology: Public Understandings,* edited by M. J. Reiss, F. Watts, & H. Wiseman, 17–32. Hackensack, NJ: World Scientific.

Lieberman, D. E. 2013. *The Story of the Human Body: Evolution, Health, and Disease.* New York: Vintage.

Lim, C., C. Barrio, M. Hernandez, A. Barragán, & J. S. Brekke. 2017. "Recovery from schizophrenia in community-based psychosocial rehabilitation settings: Rates and predictors." *Research on Social Work Practice* 27, no. 5: 538–551.

Lineweaver, C. H., P. W. Davies, & M. Ruse, eds. 2013. *Complexity and the Arrow of Time.* Cambridge: Cambridge University Press.

London Fire Brigade. 2017. "'There is no such thing as a Fireman' says Brigade Chief." https://www.london-fire.gov.uk/news/2017-news/there-is-no-such-thing-as-a-fireman-says-brigade-chief/.

Lovelock, J. E. 1979. *Gaia: A New Look at Life on Earth.* Oxford: Oxford University Press.

Lovelock, J. E. 2000. *Homage to Gaia.* Oxford: Oxford University Press.

Lovelock, J. E. 2006. *The Revenge of Gaia: Earth's Climate Crisis and the Fate of Humanity.* New York: Basic Books.

Lovelock, J. E., & L. Margulis. 1974a. "Homeostatic tendencies of the Earth's atmosphere." *Origins of Life* 5, no. 1: 93–103.

Lovelock, J. E., & L. Margulis. 1974b. "Atmospheric homeostasis by and for the biosphere: The Gaia Hypothesis." *Tellus* 26: 1–10.

Ludwig, D. S., W. C. Willett, J. S. Volek, & M. L. Neuhouser. 2018. "Dietary fat: From foe to friend?" *Science* 362, no. 6416: 764–770.

Lurie, E. 1960. *Louis Agassiz: A Life in Science.* Chicago: University of Chicago Press.

Luzzatto, L. 2012. "Sickle cell anaemia and malaria." *Mediterranean Journal of Hematology and Infectious Diseases* 4, no. 1: e2012065.

Lyell, C. 1830–1833. *Principles of Geology: Being an Attempt to Explain the Former Changes in the Earth's Surface by Reference to Causes now in Operation.* London: John Murray.

MacArthur, R. H., & E. O. Wilson. 1967. *The Theory of Island Biogeography.* Princeton, NJ: Princeton University Press.

Maguire, E. A., D. G. Gadian, I. S. Johnsrude, C. D. Good, J. Ashburner, R. S. J. Frackowiak, & C. D. Frith. 2000. "Navigation-related structural change in the hippocampi of taxi drivers." *Proceedings of the National Academy of Sciences* 97, no. 8: 4398–4403.

Malthus, T. R. [1826] 1914. *An Essay on the Principle of Population.* 6th ed. London: Everyman.

Marcus, L. 2014. "Why you shouldn't share those emotional 'Deaf person hears for the first time' videos." *The Atlantic,* March 28, 2014. https://www.theatlantic.com/politics/archive/2014/03/why-you-shouldnt-share-those-emotional-deaf-person-hears-for-the-first-time-videos/359850/.

Marks, J. 2007. "Long shadow of Linnaeus's human taxonomy." *Nature* 447: 28.

Martin, E. 1991. "The egg and the sperm: How science has constructed a romance based on stereotypical male–female roles." *Signs: Journal of Women in Culture and Society* 16: 485–501.

Mathez, E. A., & J. E. Smerdon. 2018. *Climate Change: The Science of Global Warming and Our Energy Future.* New York: Columbia University Press.

Maynard Smith, J. 1964. "Group selection and kin selection." *Nature* 201: 1145–1147.

Mayne, A. J. 1999. *From Politics Past to Politics Future: An Integrated Analysis of Current Emergent Paradigms.* Westport, CT: Praeger.

Mayo Clinic. 2020. "Schizophrenia: Treatment." https://www.mayoclinic.org/diseases-conditions/schizophrenia/diagnosis-treatment/drc-20354449.

Mayr, E. 1942. *Systematics and the Origin of Species.* New York: Columbia University Press.

Mayr, E. 1961. "Cause and effect in biology." *Science* 134, no. 3489: 1501–1506.

Mayr, E. 1969. "Commentary." *Journal of the History of Biology* 2, no. 1: 123–128.

McGrath, A. 2021. "A *Consilience of Equal Regard*: Stephen Jay Gould on the relation of science and religion." *Zygon* 56: 547–565.

Medawar, P. B. 1961. "Review of the phenomenon of man." *Mind* 70: 99–106.

MedicineNet. 2021. "Obesity." https://www.medicinenet.com/obesity_weight_loss/article.htm.

Mercedes. 2021. "Example of Ancestry DNA Results." https://whoareyoumadeof.com/blog/example-of-ancestry-dna-results/.

Merchant, C. 1980. *The Death of Nature: Women, Ecology, and the Scientific Revolution: A Feminist Reappraisal of the Scientific Revolution.* Scranton, PA: HarperCollins.

Miles, M., & V. Shiva. 2014. *Ecofeminism (Critique. Influence. Change.).* London: Zed Books.

Mill, J. S. [1843] 1974. *A System of Logic Ratiocinative and Inductive.* Edited by J. M. Robson. Toronto: University of Toronto Press.

Mitman, G. 1992. *The State of Nature: Ecology, Community, and American Social Thought, 1900–1950.* Chicago: University of Chicago Press.

Moore, A. 1890. "The Christian doctrine of God." In *Lux Mundi,* edited by C. Gore, 57–109. London: John Murray.

Moore, G. E. 1903. *Principia Ethica.* Cambridge: Cambridge University Press.

Morgan, S. R. 1990. "Schelling and the origins of his Naturphilosophie." In *Romanticism and the Sciences,* edited by A. Cunnignham & N. Jardine, 25–37. Cambridge: Cambridge University Press.

Mowat, A. 2017. "Why does cystic fibrosis display the prevalence and distribution observed in human populations?" *Current Pediatric Research* 21, no. 1: 164–171.

Muir, J. 1966. *John of the Mountains: The Unpublished Journals of John Muir.* Edited by L. M. Wolfe. Madison: University of Wisconsin Press.

Müller, G. B. 2017. "Why an extended evolutionary synthesis is necessary." *Interface Focus* 7, no. 5. https://royalsocietypublishing.org/doi/10.1098/rsfs.2017.0015.

Murray, C. 2020. *Human Diversity: The Biology of Gender, Race, and Class.* New York: Twelve.

Mussett, S. 2021. "Simone de Beauvoir (1908–1986)." https://iep.utm.edu/beauvoir/#H1.

Næss, A. 1986a. "The deep ecological movement: Some philosophical aspects." In *Deep Ecology in the 21st Century,* edited by G. Sessions, 64–84, Boston: Shambhala.

Næss, A. 1986b. "Self realization: An ecological approach to being in the world." *Deep Ecology in the 21st Century,* edited by G. Sessions, 225–239. Boston: Shambhala.

Nagel, T. 2012. *Mind and Cosmos: Why the Materialist Neo-Darwinian Conception of Nature Is Almost Certainly False.* New York: Oxford University Press.

Najibi, A. 2020. "Racial discrimination in face recognition technology." https://sitn.hms.harvard.edu/flash/2020/racial-discrimination-in-face-recognition-technology/.

National Centre for Complementary and Integrative Health. 2021. "Ayurvedic medicine: In depth." https://www.nccih.nih.gov/health/ayurvedic-medicine-in-depth.

National Institute of Mental Health. 2021. "RAISE questions and answers: Questions and answers about psychosis." https://www.nimh.nih.gov/health/topics/schizophrenia/raise/raise-questions-and-answers.shtml#3.

National Institute on Deafness and Other Communication Disorders. 2017. "Cochlear Implants." https://www.nidcd.nih.gov/health/cochlear-implants.

Navin, M. C., J. A. Wasserman, M. Ahmad, & S. Bies. 2019. "Vaccine education, reasons for refusal, and vaccination behavior." *American Journal of Preventive Medicine* 56, no. 3: 359–367.

Naydler, J., ed. 1996. *Goethe on Science: An Anthology of Goethe's Scientific Writings.* Edinburgh: Floris Books.

NCD Risk Factor Collaboration. 2016. "Trends in adult body-mass index in 200 countries from 1975 to 2014: A pooled analysis of 1698 population-based measurement studies with 19·2 million participants." *Lancet* 387: 1377–1396.

Newman, J. H. 1973. *The Letters and Diaries of John Henry Newman, XXV.* Edited by C. S. Dessain & T. Gornall. Oxford: Clarendon Press.

NHS. 2019a. "Treatment—Sickle cell disease." https://www.nhs.uk/conditions/sickle-cell-disease/treatment/.

NHS. 2019b. "Causes—Schizophrenia." https://www.nhs.uk/conditions/schizophrenia/causes/.

NHS. 2019c. "Acupuncture." https://www.nhs.uk/conditions/acupuncture/.

NHS Inform. 2021. "Obesity." https://www.nhsinform.scot/illnesses-and-conditions/nutritional/obesity.

Nicholson, D. J. 2014. "The return of the organism as a fundamental explanatory concept in biology." *Philosophy Compass* 9: 347–359.

Nicholson, D. J., and R. Gawne. 2014. "Rethinking Woodger's legacy in the philosophy of biology." *Journal of the History of Biology* 47: 243–292.

Nicholson, D. J., and R. Gawne. 2015. "Neither logical empiricism nor vitalism, but organicism: What the philosophy of biology was." *History and Philosophy of the Life Sciences* 37: 345–381.

Noble, D. 2006. *The Music of Life.* Oxford: Oxford University Press.

Noble, D. 2017. *Dance to the Tune of Life: Biological Relativity.* Cambridge: Cambridge University Press.

Noll, M. 2002. *America's God: From Jonathan Edwards to Abraham Lincoln.* New York: Oxford University Press.
Noonan, R. 2020. "We've known for over a century that our environment shapes our health, so why are we still blaming unhealthy lifestyles?" *The Conversation,* September 9, 2020. https://theconversation.com/weve-known-for-over-a-century-that-our-environment-shapes-our-health-so-why-are-we-still-blaming-unhealthy-lifestyles-145597?utm_medium=email&utm_campaign=Latest%20from%20The%20Conversation%20for%20September%2010%202020%20-%201727216692&utm_content=Latest%20from%20The%20Conversation%20for%20September%2010%202020%20-%201727216692+CID_783da2bed13f54638969984e87b367ca&utm_source=campaign_monitor_uk&utm_term=Weve%20known%20for%20over%20a%20century%20that%20our%20environment%20shapes%20our%20health%20so%20why%20are%20we%20still%20blaming%20unhealthy%20lifestyles.
Nordbakke, S. 2019. "Children's out-of-home leisure activities: Changes during the last decade in Norway." *Children's Geographies* 17, no. 3: 347–360.
Norwood, V. L. 1987. "The nature of knowing: Rachel Carson and the American Environment." *Signs* 12: 740–760.
Novartis. 2019. "New Novartis medicine Adakveo® (crizanlizumab) approved by FDA to reduce frequency of pain crises in individuals living with sickle cell disease." https://www.novartis.com/news/media-releases/new-novartis-medicine-adakveo-crizanlizumab-approved-fda-reduce-frequency-pain-crises-individuals-living-sickle-cell-disease.
Numbers, R. L. 2006. *The Creationists: From Scientific Creationism to Intelligent Design.* Standard edition. Cambridge, MA: Harvard University Press.
Nursing and Midwifery Council. 2016. "Quantity of midwives registered with the NMC that are male." https://web.archive.org/web/20161008111419/https://www.whatdotheyknow.com/request/quantity_of_midwives_registered.
Nyhan, B., J. Reifler, S. Richey, & G. L. Freed. 2014. "Effective messages in vaccine promotion: A randomized trial." *Pediatrics* 133, no. 4: e835–e842.
Obama, B. 2019. https://twitter.com/thehill/status/1189409128587956226.
O'Brien, B. 2019. "The principle of dependent origination in Buddhism." https://www.learnreligions.com/dependent-origination-meaning-449723.
O'Connell, G. 2017. "Will Pope Francis remove the Vatican's 'warning' from Teilhard de Chardin's writings?" *America,* November 21, 2017. https://www.americamagazine.org/faith/2017/11/21/will-pope-francis-remove-vaticans-warning-teilhard-de-chardins-writings.
O'Connell, J., & M. Ruse. 2021. *Social Darwinism.* Cambridge: Cambridge University Press.
Office for National Statistics. 2021. "Ethnic group, national identity and religion." https://www.ons.gov.uk/methodology/classificationsandstandards/measuringequality/ethnicgroupnationalidentityandreligion.
Oliver, M., & C. Barnes. 2010. "Disability studies, disabled people and the struggle for inclusion." *British Journal of Sociology of Education* 31, no. 5: 547–560.

O'Neill, C. 2003. "d or D? Who's deaf and who's Deaf?" http://www.bbc.co.uk/ouch/opinion/d_or_d_whos_deaf_and_whos_deaf.shtml.

Open Society Foundations. 2019. "What are intersex rights?" https://www.opensocietyfoundations.org/explainers/what-are-intersex-rights.

Palmer, L. 2020. "Why green spaces, walkable neighbourhoods and life-enhancing buildings can all help in the fight against dementia." *The Conversation,* September 25, 2020. https://theconversation.com/why-green-spaces-walkable-neighbourhoods-and-life-enhancing-buildings-can-all-help-in-the-fight-against-dementia-146712?utm_medium=email&utm_campaign=Latest%20from%20The%20Conversation%20for%20September%2028%202020%20-%201743616882&utm_content=Latest%20from%20The%20Conversation%20for%20September%2028%202020%20-%201743616882+CID_ec3a8c6e6045e46f1c352964aa200778&utm_source=campaign_monitor_uk&utm_term=Why%20green%20spaces%20walkable%20neighbourhoods%20and%20life-enhancing%20buildings%20can%20all%20help%20in%20the%20fight%20against%20dementia.

Parker, G. A. 1978. "Searching for mates." In *Behavioural Ecology: An Evolutionary Approach.* Sunderland, MA: Sinauer.

Partridge, E. A., M. G. Davey, M. A. Hornick, P. E. McGovern, A. Y. Mejaddam, J. D. Vrecenak, C. Mesas-Burgos, A. Olive, R. C. Caskey, T. R. Weiland, J. Han, A. J. Schupper, J. T. Connelly, K. C. Dysart, J. Rychik, H. L. Hedrick, W. H. Peranteau, & A. W. Flake. 2017. "An extra-uterine system to physiologically support the extreme premature lamb." *Nature Communications* 8: 15112. https://doi.org/10.1038/ncomms15112.

Peacocke, A. R. 1986. *God and the New Biology.* London: Dent.

Peacocke, A. R. 2001. *Paths from Science towards God: The End of all our Exploring.* Oxford: Oneworld.

Peixoto, M. C., J. Spratley, G. Oliveira, J. Martins, J. Bastosa, & C. Ribeiroa. 2013. "Effectiveness of cochlear implants in children: Long term results." *International Journal of Pediatric Otorhinolaryngology* 77, no. 4: 462–468.

Pelčić, G., S. Karačić, G. L. Mikirtichan, O. I. Kubar, F. J. Leavitt, M. Cheng-Tek Tai, N. Morishita, S. Vuletić, & L. Tomašević. 2016. "Religious exception for vaccination or religious excuses for avoiding vaccination." *Croatian Medical Journal* 57, no. 5: 516–521.

Peterson, E. L. 2016. *The Life Organic: The Theoretical Biology Club and the Roots of Epigenetics.* Pittsburgh: University of Pittsburgh Press.

Pfeiffer, E. E. 1958. "Do we know what we are doing? DDT spray programs—their values and dangers." *Bio-Dynamics* 45: 2–40.

Pilipinas, K. M., P. M. Layosa, K. Acharya, R. F. Quijano, A. Sampaguita, W. R. Pelegrina, & P. Aurelio Z. dela Cruz. 2007. *The Great Rice Robbery.* Penang, Malaysia: Pesticide Action Network.

Plater, R. 2020. "First person treated for sickle cell disease with CRISPR is doing well." *Healthline,* July 6, 2020. https://www.healthline.com/health-news/first-person-treated-for-sickle-cell-disease-with-crispr-is-doing-well.

Plato. 1941. *The Republic.* Translated by F. Cornford. Oxford: Oxford University Press.

Price, L. 1954. *Dialogues of Whitehead*. Boston: Little, Brown.

Prieto, G. I. 2021. Review of: Kostas Kampourakis and Tobias Uller (Eds.): Philosophy of Science for Biologists. *Journal for the General Philosophy of Science* 52: 613–615.

Prince of Wales, T. Juniper, & I. Skelly. 2010. *Harmony: A New Way of Looking at Our World*. New York: Harper Collins.

Provine, W. B. 1971. *The Origins of Theoretical Population Genetics*. Chicago: University of Chicago Press.

Provine, W. B. 1986. *Sewall Wright and Evolutionary Biology*. Chicago: University of Chicago Press.

Rackham, O. 2015. *Woodlands*. Glasgow, UK: William Collins.

Rahner, K. 1966. "Christology within an evolutionary view of the world." In *Theological Investigations. V. Later Writings*, 157–192. Baltimore: Helicon Press.

Razak, A. 1990. "Toward a womanist analysis of birth." In *Reweaving the World: The Emergence of Ecofeminism*, edited by I. Diamond & G. F. Orenstein, 165–172. San Francisco: Sierra Club.

Regis, E. 2019. *Golden Rice: The Imperiled Birth of a GMO Superfood*. Baltimore: Johns Hopkins University Press.

Reiss, M. J. 2019. "Evolution education: Treating evolution as a sensitive rather than a controversial issue." *Ethics and Education* 14, no. 3: 351–366.

Reiss, M. J. 2020. "Food and nutrition." In *Rethinking Biology: Public Understandings*, edited by M. J. Reiss, F. Watts, & H. Wiseman, 213–230. Hackensack, NJ: World Scientific.

Reiss, M. J., & S. D. Tunnicliffe. 2001. "What sorts of worlds do we live in nowadays? Teaching biology in a post-modern age." *Journal of Biological Education* 35: 125–129.

Reiss, M. J., F. Watts, & H. Wiseman, eds. 2020. *Rethinking Biology: Public Understandings*. Hackensack, NJ: World Scientific.

Repcheck, J. 2003. *The Man Who Found Time: James Hutton and the Discovery of Earth's Antiquity*. New York: Perseus.

Richards, R. J. 2003. *The Romantic Conception of Life: Science and Philosophy in the Age of Goethe*. Chicago: University of Chicago Press.

Richards, R. J. 2008. *The Tragic Sense of Life: Ernst Haeckel and the Struggle over Evolutionary Thought*. Chicago: University of Chicago Press.

Ridgwell, H. 2013. "Pope John Paul II to become saint after miracle approved." https://www.voanews.com/europe/pope-john-paul-ii-become-saint-after-miracle-approved.

Riedel, S. 2005. "Edward Jenner and the history of smallpox and vaccination." *Baylor University Medical Center Proceedings* 18: 21–25.

Rippon, G. 2019. *The Gendered Brain: The New Neuroscience that Shatters the Myth of the Female Brain*. London: Bodley Head.

Rivolo, M. 2018. "SICKLE as a holistic treatment approach to sickle cell disease related ulcers." *British Journal of Nursing* 27, no. 20. https://doi.org/10.12968/bjon.2018.27.Sup20.S6.

Roe, J., & P. Aspinall. 2011. "The restorative benefits of walking in urban and rural settings in adults with good and poor mental health." *Health & Place* 17: 103–113.

Rogers, P. 2001. *Song of the World Becoming: New and Collected Poems 1981–2001.* Minneapolis, MN: Milkweed.

Rolston III, H. 1988. *Environmental Ethics.* Philadelphia: Temple University Press.

Rose, H., & S. Rose. 2013. *Genes, Cells and Brains: The Promethean Promises of the New Biology.* London: Verso.

Rose, H., & S. Rose. 2016. *Can Neuroscience Change Our Minds?* Cambridge: Polity Press.

Rosenberg, N. A., J. K. Pritchard, J. L. Weber, H. M. Cann, K. K. Kidd, L. A. Zhivotovsky, & M. W. Feldman. 2002. "Genetic structure of human populations." *Science* 298, no. 5602: 2381–2385.

Roth, L. 2021. The colour balance project. http://colourbalance.lornaroth.com/texts/.

Roth, W. D., & B. Ivemark. 2018. "Genetic options: The impact of genetic ancestry testing on consumers' racial and ethnic identities." *American Journal of Sociology* 124, no. 1: 150–184.

Rothman, K. 2008. "BMI-related errors in the measurement of obesity." *International Journal of Obesity* 32: S56–S59.

Rudwick, M. J. S. 1972. *The Meaning of Fossils.* New York: Science History.

Rudwick, M. J. S. 2005. *Bursting the Limits of Time.* Chicago: University of Chicago Press.

Rund, B. R. 2018. "A review of factors associated with severe violence in schizophrenia." *Nordic Journal of Psychiatry* 72, no. 8: 561–571.

Ruse, M. 1973. *The Philosophy of Biology.* London: Hutchinson.

Ruse, M. 1975. "Darwin's debt to philosophy: An examination of the influence of the philosophical ideas of John F. W. Herschel and William Whewell on the development of Charles Darwin's theory of evolution." *Studies in History and Philosophy of Science* 6: 159–181.

Ruse, M. 1981. "What kind of revolution occurred in geology?" *PSA: Proceedings of the Biennial Meeting of the Philosophy of Science Association* 2: 240–273.

Ruse, M. 1982. *Darwinism Defended: A Guide to the Evolution Controversies.* Reading, MA: Benjamin / Cummings.

Ruse, M., ed. 1988. *But is it Science? The Philosophical Question in the Creation/ Evolution Controversy.* Buffalo, NY: Prometheus.

Ruse, M. 1996. *Monad to Man: The Concept of Progress in Evolutionary Biology.* Cambridge, MA: Harvard University Press.

Ruse, M. 2004. "Adaptive landscapes and dynamic equilibrium: The Spencerian contribution to twentieth-century American evolutionary biology." In *Darwinian Heresies,* edited by A. Lustig, R. J. Richards, & M. Ruse, 131–150. Cambridge: Cambridge University Press.

Ruse, M. 2010. *Science and Spirituality: Making Room for Faith in the Age of Science.* Cambridge: Cambridge University Press.

Ruse, M. 2012. *The Philosophy of Human Evolution.* Cambridge: Cambridge University Press.

Ruse, M. 2013. *The Gaia Hypothesis: Science on a Pagan Planet.* Chicago: University of Chicago Press.

Ruse, M. 2015. *Atheism: What Everyone Needs to Know*. Oxford: Oxford University Press.

Ruse, M. 2017. *On Purpose*. Princeton, NJ: Princeton University Press.

Ruse, M. 2021. "The scientific revolution." In *The Cambridge History of Atheism*, edited by S. Bullivant & M. Ruse, 1, 258–277. Cambridge: Cambridge University Press.

Ruse, M., & E. O. Wilson. 1985. "The evolution of morality." *New Scientist* 1478: 108–128.

Ruse, M., & E. O. Wilson. 1986. "Moral philosophy as applied science." *Philosophy* 61: 173–192.

Russell, F. 2021. "Covid vaccinations—Advice to schools." https://www.steinerwaldorf.org/covid-vaccinations-advice-to-schools/.

Rutherford, A. 2020 / 2021. *How to Argue with a Racist: History, Science, Race and Reality*. London: Weidenfeld & Nicolson.

Sacks, O. 1989. *Seeing Voices: A Journey into the World of the Deaf*. Berkeley: University of California Press.

Sandall, J., H. Soltani, S. Gates, A. Shennan, & D. Devane. 2016. "Midwife-led continuity models versus other models of care for childbearing women." *Cochrane Database Systematic Reviews* 4: CD004667.

Sandberg, D. E., V. Pasterski, & N. Callens. 2017. "Introduction to the Special Section: Disorders of sex development." *Journal of Pediatric Psychology* 42, no. 5: 487–495.

Sanders, D. L., & D. Jenkins. 2018. "Plant biology." In *Teaching Biology in Schools: Global Research, Issues, and Trends*, edited by K. Kampourakis & M. J. Reiss, 124–138. New York: Routledge.

Sanders, L. 2020. "The difference between what Republicans and Democrats believe to be true about COVID-19." *YouGov*, May 26, 2022. https://today.yougov.com/topics/politics/articles-reports/2020/05/26/republicans-democrats-misinformation.

Sax, L. 2002. "How common is intersex? A response to Anne Fausto-Sterling." *The Journal of Sex Research* 39, no. 3: 174–178.

Schelling, F. W. J. [1797] 1988. *Ideas for a Philosophy of Nature*. Cambridge: Cambridge University Press.

Schelling, F. W. J. [1833–1834] 1994. *On the History of Modern Philosophy*. Cambridge: Cambridge University Press.

Schiff, M., M. Duyme, A. Dumaret, & S. Tomkiewicz. 1982. "How much could we boost scholastic achievement and IQ scores? A direct answer from a French adoption study." *Cognition* 12, no. 2: 165–196.

Schulster, M., A. M. Bernie, & R. Ramasamy. 2016. "The role of estradiol in male reproductive function." *Asian Journal of Andrology* 18, no. 3: 435–440.

Scott, C., J. Sutherland, & A. Taylor. 2018. "Affordability of the UK's Eatwell Guide." https://foodfoundation.org.uk/wp-content/uploads/2018/10/Affordability-of-the-Eatwell-Guide_Final_Web-Version.pdf.

Sedley, D. 2008. *Creationism and its Critics in Antiquity*. Berkeley: University of California Press.

Seidelmann, S. B., B. Claggett, S. Cheng, M. Henglin, A. Shah, L. M. Steffen, A. R. Folsom, E. B. Rimm, W. C. Willett, & S. D. Solomon. 2018. "Dietary carbohydrate intake and mortality: A prospective cohort study and meta-analysis." *Lancet Public Health* 3: e419–e428.

Shakespeare, T. 2006. "The social model of disability." In *The Disability Studies Reader,* edited by L. J. Davies, 2nd ed., 197–204. New York: Routledge.

Sharma, K. 2015. *Interdependence: Biology and Beyond.* New York: Fordham University Press.

Sheppard, P. M. 1958. *Natural Selection and Heredity.* London: Hutchinson.

Shiva, V. 2001. "The 'Golden Rice' hoax: When public relations replaces science. Seed Freedom. https://seedfreedom.info/campaign/the-golden-rice-hoax/.

Shrier, A. 2020. *Irreversible Damage: Teenage Girls and the Transgender Craze.* Washington, DC: Regnery.

Signorile, A. L., P. W. Lurz, J. Wang, D. C. Reuman, & C. Carbone. 2016. "Mixture or mosaic? Genetic patterns in UK grey squirrels support a human-mediated 'long-jump' invasion mechanism." *Diversity and Distributions* 22, no. 5: 566–577.

SimilarWeb. 2021. "Top sites ranking for health in the world." https://www.similarweb.com/top-websites/category/health/.

Simpson, G. G. 1944. *Tempo and Mode in Evolution.* New York: Columbia University Press.

Sini, R. 2016. "'Three black teenagers' Google search sparks Twitter row." *BBC News,* June 6, 2016. https://www.bbc.co.uk/news/world-us-canada-36487495.

Smith, A. [1776] 1937. *The Wealth of Nations.* New York: Modern Library.

Smocovitis, V. B. 1999. "The 1959 Darwin centennial celebration in America." *Osiris* 14: 274–323.

Smuts, J. C. 1926. *Holism and Evolution.* London: Macmillan.

Sober, E., & Wilson, D. S. 1998. *Unto Others: The Evolution and Psychology of Unselfish Behavior.* Cambridge, MA: Harvard University Press.

Sobo, E. J. 2015. "Social cultivation of vaccine refusal and delay among Waldorf (Steiner) school parents." *Medical Anthropology Quarterly* 29, no. 3: 381–399.

Sotos, J. G. 2003. "Taft and Pickwick: Sleep apnea in the White House." *Chest* 124, no. 3: 1133–1142.

Southgate, C. 2008. *The Groaning of Creation: God, Evolution, and the Problem of Evil.* Louisville, KY: Westminster John Knox Press.

Specter, M. 2014. "Seeds of doubt: An activist's controversial crusade against genetically modified crops." *New Yorker.* https://www.newyorker.com/magazine/2014/08/25/seeds-of-doubt.

Spencer, H. 1851. *Social Statics: Or, the Conditions Essential to Human Happiness Specified, and the First of Them Developed.* London: Chapman.

Spencer, H. 1852a. "A theory of population, deduced from the general law of animal fertility." *Westminster Review* 1: 468–501.

Spencer, H. [1852b] 1868. "The development hypothesis. The Leader." In *Essays: Scientific, Political and Speculative,* 377–383. London: Williams and Norgate.

Spencer, H. [1857] 1868. "Progress: Its law and cause." *Westminster Review* LXVII: 244–267. Reprinted in Essays Scientific, Political and Speculative.

Spencer, H. 1860. "The social organism." *Westminster Review* LXXIII: 90–121.
Spencer, H. 1862. *First Principles.* London: Williams and Norgate.
Spencer, H. 1879. *The Data of Ethics.* London: Williams and Norgate.
Spretnak, C. 1989. "Toward an ecofeminist spirituality." In *Healing the Wounds: The Promise of Ecofeminism,* edited by J. Plant, 127–132. Philadelphia: New Society.
Stanley, S., & S. Shwetha. 2006. "Integrated psychosocial intervention in schizophrenia: Implications for patients and their caregivers." *International Journal of Psychosocial Rehabilitation* 10, no. 2: 113–128.
Stanton, R. 2021. *The Disneyfication of Animals.* Cham, Switzerland: Palgrave Macmillan.
Stebbins, G. L. 1950. *Variation and Evolution in Plants.* New York: Columbia University Press.
Steiner, R. 1924. *Agricultural Course: Birth of the Biodynamic Method.* Forest Row, UK: Rudolf Steiner Press.
Sterelny, K. 2012. *The Evolved Apprentice: How Evolution Made Humans Unique.* Cambridge, MA: MIT Press.
Stock, K. 2021. *Material Girls: Why Reality Matters for Feminism.* London: Fleet.
Sturm, V. E., S. Datta, A. R. K. Roy, I. J. Sible, E. L. Kosik, C. R. Veziris, T. E. Chow, N. A. Morris, J. Neuhaus, J. H. Kramer, B. L. Miller, S. R. Holley, & D. Keltner. 2020. "Big smile, small self: Awe walks promote prosocial positive emotions in older adults." *Emotion* 22, no. 5: 1044–1058.
Suissa, J., & A. Sullivan. 2021. "The gender wars, academic freedom and education." *Journal of Philosophy of Education* 55, no. 1: 55–82.
Sussman, R. W. 2014. *The Myth of Race: The Troubling Persistence of an Unscientific Idea.* Cambridge, MA: Harvard University Press.
Syme, K. L., & E. H. Hagen. 2020. "Mental health is biological health: Why tackling 'diseases of the mind' is an imperative for biological anthropology in the 21st century." *Yearbook of Physical Anthropology* 171: 87–117.
Taylor, P. W. 1981. "The ethics of respect for nature." *Environmental Ethics* 3, no. 3: 197–218.
Teilhard de Chardin, P. 1955. *The Phenomenon of Man.* London: Collins.
The New York Times. 2020. "Alexander Walker." https://www.nytimes.com/interactive/2015/opinion/transgender-today/stories/alexander-walker?mtrref=www.nytimes.com&gwh=B70D5B28EEC2AEC3BD37C584D07D6586&gwt=pay&assetType=REGIWALL.
Thomas, C. 1991. "Sexual assault: Issues for aboriginal women." https://pdfs.semanticscholar.org/53e9/3849f0e99fcb584de03c8e52b9020cc63f10.pdf?_ga=2.51533225.1453723003.1598797077-1362316038.1598797077.
Thomas, P. 2020. "No smoke without ire." *Radio Times,* October 3–9, 2020, 142.
Thompson, D. W. 1917. *On Growth and Form.* Cambridge: Cambridge University Press.
Thoreau, H. 1854. *Walden; or Life in the Woods.* Boston: Ticknor and Fields.

Topping, A. 2020. "Want gender equality? Then fight for fathers' rights to shared parental leave." *The Guardian,* February 11, 2020. https://www.theguardian.com/commentisfree/2020/feb/11/gender-pay-gap-shared-parental-leave-finland.
Trivers, R. L. 1971. "The evolution of reciprocal altruism." *Quarterly Review of Biology* 46: 35–57.
UK Disability History Month. 2017. "August Natterer, 1868–1933." https://ukdhm.org/august-natterer-1868-to-1933-delerium-anxiety-and-hallucinations/.
Ulijaszek, S. J., & A. K. McLennan. 2016. "Framing obesity in UK policy from the Blair years, 1997–2015: The persistence of individualistic approaches despite overwhelming evidence of societal and economic factors, and the need for collective responsibility." *Obesity Reviews* 17: 397–411.
United Nations Children's Fund. 2016. *Female Genital Mutilation/Cutting: A Global Concern.* New York: UNICEF.
University of California Museum of Paleontology. 2019. "Understanding evolution: One route to lighter skin, two different continents." https://evolution.berkeley.edu/evolibrary/news/190122_allele.
University of Exeter. 2017. "Clownfish males become fierce females if their 'wife' is eaten." http://www.exeter.ac.uk/news/featurednews/title_594884_en.html.
Vallejos, M., O. M. Cesoni, R. Farinola, M. S. Bertone, & C. R. Prokopez. 2017. "Adverse childhood experiences among men with schizophrenia." *Psychiatric Quarterly* 88, no. 4: 665–673.
Van Dyke, F., D. C. Mahan, J. K. Sheldon, & R. H. Brand. 1996. *Redeeming Creation: The Biblical Basis for Environmental Stewardship.* Downers Grove, IL: InterVarsity Press.
van Leeuwen, C., & M. Herschbach. 2020. "Editors' note." *Philosophical Psychology* 33, no. 1: 148–150.
VanderKruik, R., M. Barreix, D. Chou, T. Allen, L. Say, L. S. Cohen, & Maternal Morbidity Working Group. 2017. "The global prevalence of postpartum psychosis: A systematic review." *BMC Psychiatry* 17, no. 1: 272.
Veenendaal, M. V., R. C. Painter, S. R. de Rooij, P. M. Bossuyt, J. A. van der Post, P. D. Gluckman, M. A. Hanson, & T. J. Roseboom. 2013. "Transgenerational effects of prenatal exposure to the 1944–45 Dutch famine." *BJOG: An International Journal of Obstetrics & Gynaecology* 120, no. 5: 548–553.
Veile, A., & V. Miller. 2019. "Duration of breast feeding in ancestral environments." In *Encyclopedia of Evolutionary Psychological Science,* edited by T. Shackelford & V. Weekes-Shackelford, Cham, Switzerland: Springer.
Ventegodt, S., I. Kandel, & J. Merrick. 2007. "A short history of clinical holistic medicine." *The Scientific World Journal* 7: 1622–1630.
Voit, E. O. 2016 *The Inner Workings of Life: Vignettes in Systems Biology.* Cambridge: Cambridge University Press.
Waddington, C. H. 1953. "Genetic assimilation of an acquired character." *Evolution* 7: 118–126.
Waddington, C. H. 1957. *The Strategy of the Genes.* London: Allen and Unwin.
Walajahi, H., D. R. Wilson, & S. C. Hull. 2019. "Constructing identities: The implications of DTC ancestry testing for tribal communities." *Genetics in Medicine* 21: 1744–1750.

Watson, A. J., & J. E. Lovelock. 1983. "Biological homeostasis of the global environment: The parable of Daisyworld." *Tellus, Series B: Chemical and Physical Meterology* 35, no. 4: 284–289.
Watson, J. D., & F. H. C. Crick. 1953. "Molecular structure of nucleic acids; A structure for deoxyribose nucleic acid." *Nature* 171, no. 4356: 737–738.
Watts, F., & M. J. Reiss. 2017. "Holistic biology: What it is and why it matters." *Zygon* 52: 419–441.
Weart, S. R. 2008. *The Discovery of Global Warming*. Cambridge, MA: Harvard University Press.
WebMD. 2020. "What is holistic medicine?" https://www.webmd.com/balance/guide/what-is-holistic-medicine#1.
Weinberg, S. 1977. *The First Three Minutes: A Modern View of the Origin of the Universe*. New York: Basic Books.
Wesseler, J., & D. Zilberman. 2014. "The economic power of the Golden Rice opposition." *Environment and Development Economics* 19: 724–742.
Wheeler, W. M. 1939. *Essays in Philosophical Biology*. Cambridge, MA: Harvard University Press.
Whewell, W. 1837. *The History of the Inductive Sciences*. 3 vols. London: Parker.
Whewell, W. 1840. *The Philosophy of the Inductive Sciences*. 2 vols. London: Parker.
White, A. D. 1896. *History of the Warfare of Science with Theology in Christendom*. New York: Appleton.
Whitehead, A. N. 1926. *Science and the Modern World*. Cambridge: Cambridge University Press.
Whitehead, A. N. [1929] 1978. *Process and Reality: An Essay in Cosmology*. New York: Free Press.
Whitehead, A. N. 1938. "Nature alive." In *Modes of Thought*, 202–232. New York: Macmillan.
Whitehead, A. N. 1960. *Religion in the Making*. New York: Meridian.
Wiesemann, C., S. Ude-Koeller, G. H. Sinnecker, & U. Thyen. 2010. "Ethical principles and recommendations for the medical management of differences of sex development (DSD) / intersex in children and adolescents." *European Journal of Pediatrics* 169, no. 6: 671–679.
Wikipedia. 2021a. "Alternative medicine." https://en.wikipedia.org/wiki/Alternative_medicine.
Wikipedia. 2021b. "Killing of George Floyd." https://en.wikipedia.org/wiki/Killing_of_George_Floyd.
Williams, B. A., H. McCartney, E. Adams, A. M. Devlin, J. Singer, S. Vercauteren, J. K. Wu, & C. D. Karakochuk. 2020. "Folic acid supplementation in children with sickle cell disease: Study protocol for a double-blind randomized cross-over trial." *Trials* 21: 593.
Williams, G. 2010. *Angel of Death*. Basingstoke, UK: Palgrave Macmillan.
Willick, M. S. 2001. "Psychoanalysis and schizophrenia: A cautionary tale." *Journal of the American Psychoanalytic Association* 49, no. 1: 27–56.
Willyard, C. 2018. "New human gene tally reignites debate." *Nature* 558: 354–355.

Wilshaw, A. 2018. "Out of Africa hypothesis." In *The International Encyclopedia of Biological Anthropology*, edited by W. Trevathan, M. Cartmill, D. Dufour, C. Larsen, D. O'Rourke, K. Rosenberg, & K. Strier. Hoboken, NJ: Wiley.
Wilson, D. S., & E. O. Wilson. 2007. "Rethinking the theoretical foundation of sociobiology." *Quarterly Review of Biology* 82: 327–348.
Wilson, E. O. 1975. *Sociobiology: The New Synthesis*. Cambridge, MA: Harvard University Press.
Wilson, E. O. 1984. *Biophilia*. Cambridge, MA: Harvard University Press.
Wilson, E. O. 1992. *The Diversity of Life*. Cambridge, MA: Harvard University Press.
Wilson, E. O. 1994. *Naturalist*. Washington, DC: Island Books / Shearwater Books.
Winsor, M. P. 1991. *Reading the Shape of Nature: Comparative Zoology at the Agassiz Museum*. Chicago: University of Chicago Press.
Wiseman, H. 2016. *The Myth of the Moral Brain: The Limits of Moral Enhancement*. Cambridge, MA: MIT Press.
Wittgenstein, L. 1965. "A lecture on ethics." *The Philosophical Review* 74: 3–12.
Woodger, J. H. 1929. *Biological Principles*. London: Routledge.
World Health Organization. 2012. "Population-based approaches to childhood obesity prevention." https://apps.who.int/iris/bitstream/handle/10665/80149/9789241504782_eng.pdf.
World Health Organization. 2019a. "Schizophrenia." https://www.who.int/news-room/fact-sheets/detail/schizophrenia.
World Health Organization. 2019b. "Ten threats to global health in 2019." https://www.who.int/news-room/spotlight/ten-threats-to-global-health-in-2019.
Worster, D. 2008. *A Passion for Nature: The Life of John Muir*. New York: Oxford University Press.
Wren, B., J. Launer, M. J. Reiss, A. Swanepoel, & G. Music. 2019. "Can evolutionary thinking shed light on gender diversity?" *BJPsych Advances* 25, no. 6: 351–362.
Wright, S. 1931. "Evolution in Mendelian populations." *Genetics* 16: 97–159.
Wright, S. 1932. "The roles of mutation, inbreeding, crossbreeding and selection in evolution." *Proceedings of the Sixth International Congress of Genetics* 1: 356–366.
Zell-Ravenheart, O. 2004. *Grimoire for the Apprentice Wizard*. Franklin Lakes, NJ: New Page Books.
Zell-Ravenheart, O. 2009. *Green Egg Omelet: An Anthology of Art and Articles from the Legendary Pagan Journal*. Franklin Lakes, NJ: New Page Books.
Zitser, J. 2021. "Joe Biden is 'healthy, vigorous,' and fit for office on his 79th birthday, says White House doctor in post-physical report." https://www.businessinsider.com/joe-biden-is-healthy-vigorous-colonoscopy-fit-office-wh-doctor-2021-11?r=US&IR=T.

ACKNOWLEDGMENTS

We are grateful to Templeton World Charity Foundation, which funded the project "The New Biology: Implications for Philosophy, Theology and Education," on which we both worked and from which this book is one of the outputs. We are also very grateful to our Harvard University Press editor, Janice Audet, for her guidance and patience.

Michael Reiss, as always, thanks the person to whom he is married, Jenny L. Chapman, for her untiring support and great patience for all the evenings and weekends he spent typing away in the study.

Michael Ruse would like to thank William and Lucyle Werkmeister, whose bequest to the Philosophy Department at Florida State University funded his professorship and provided research funds that supported his writing of this book. As always, he thanks his wife, Lizzie, for her support. She has an undergraduate degree from the Ontario Agricultural College. Delicacy forbids him revealing her opinion on alternative methods of animal and plant care. Their cairn terriers, however, are omnivores and, so long as it is not vegan, could not care less about how their kibble was produced.

INDEX

Page numbers in *italics* refer to figures.